Information Technology for Manufacturing

A Research Agenda

Committee to Study Information Technology and Manufacturing

Computer Science and Telecommunications Board
Commission on Physical Sciences, Mathematics, and Applications

Manufacturing Studies Board
Commission on Engineering and Technical Systems

National Research Council

National Academy Press
Washington, D.C. 1995

NOTICE: The project that is the subject of this report was approved by the Governing Board of the National Research Council, whose members are drawn from the councils of the National Academy of Sciences, the National Academy of Engineering, and the Institute of Medicine. The members of the committee responsible for the report were chosen for their special competences and with regard for appropriate balance.

This report has been reviewed by a group other than the authors according to procedures approved by a Report Review Committee consisting of members of the National Academy of Sciences, the National Academy of Engineering, and the Institute of Medicine.

Support for this project was provided by the National Science Foundation (under Grant No. MIP-93/2296). Any opinions, findings, conclusions, or recommendations expressed in this material are those of the authors and do not necessarily reflect the views of the National Science Foundation.

Library of Congress Catalog Card Number 94-67789
International Standard Book Number 0-309-05179-7

Cover: The manufacturing "wheel" shown is intended to suggest the integration through information technology of manufacturing activities both among and across the four basic elements of an idealized process. The concept is discussed in Chapter 2 of this report.

Available from:

National Academy Press
2101 Constitution Avenue, NW
Box 285
Washington, DC 20055
800-624-6242
202-334-3313 (in the Washington Metropolitan Area)

B-483

iv

MANUFACTURING STUDIES BOARD

The National Academy of Sciences is a private, nonprofit, self-perpetuating society of distinguished scholars engaged in scientific and engineering research, dedicated to the furtherance of science and technology and to their use for the general welfare. Upon the authority of the charter granted to it by the Congress in 1863, the Academy has a mandate that requires it to advise the federal government on scientific and technical matters. Dr. Bruce Alberts is president of the National Academy of Sciences.

The National Academy of Engineering was established in 1964, under the charter of the National Academy of Sciences, as a parallel organization of outstanding engineers. It is autonomous in its administration and in the selection of its members, sharing with the National Academy of Sciences the responsibility for advising the federal government. The National Academy of Engineering also sponsors engineering programs aimed at meeting national needs, encourages education and research, and recognizes the superior achievements of engineers. Dr. Robert M. White is president of the National Academy of Engineering.

The Institute of Medicine was established in 1970 by the National Academy of Sciences to secure the services of eminent members of appropriate professions in the examination of policy matters pertaining to the health of the public. The Institute acts under the responsibility given to the National Academy of Sciences by its congressional charter to be an adviser to the federal government and, upon its own initiative, to identify issues of medical care, research, and education. Dr. Kenneth I. Shine is president of the Institute of Medicine.

The National Research Council was organized by the National Academy of Sciences in 1916 to associate the broad community of science and technology with the Academy's purposes of furthering knowledge and advising the federal government. Functioning in accordance with general policies determined by the Academy, the Council has become the principal operating agency of both the National Academy of Sciences and the National Academy of Engineering in providing services to the government, the public, and the scientific and engineering communities. The Council is administered jointly by both Academies and the Institute of Medicine. Dr. Bruce Alberts and Dr. Robert M. White are chairman and vice chairman, respectively, of the National Research Council.

Preface

At the request of the National Science Foundation, the National Research Council's Computer Science and Telecommunications Board (CSTB) and Manufacturing Studies Board (MSB) formed the Committee to Study Information Technology and Manufacturing in April 1993. The committee of 16 individuals from academia and industry was charged with determining the computer science and engineering research needed to support advanced manufacturing.

In preparing this report, the committee reviewed and synthesized relevant material from recent reports and initiatives, interviewed a number of researchers and practitioners in the field, and met five times to discuss the input from these sources as well as the independent observations and findings of the committee members themselves. (Contributors to this report are listed in Appendix A.) It shared its preliminary thinking with a broad community in an interim report.[1] Reviewers of that document noted a number of issues as deserving of further study that are addressed in this final report. The committee included experts from the information technology (IT) and the manufacturing domains, individuals involved in research and development as well as implementation, and individuals experienced in the manufacture of mechanical and electronic products.[2] In short,

---

[1] Computer Science and Telecommunications Board and Manufacturing Studies Board, National Research Council. 1993. *Information Technology and Manufacturing: A Preliminary Report on Research Needs.* National Academy Press, Washington, D.C.

[2] In the interests of keeping this project manageable, the committee concentrated on discrete manufacturing. It did not address continuous manufacturing (the production of substances and materials) to any significant degree.

the committee was by design highly heterogeneous, a characteristic intended to promote discussion and synergy among its members.

The committee focused on articulating a vision of IT-enabled manufacturing in the 21st century, identifying the obstacles to achieving the vision, and identifying research topics that address the obstacles. Its deliberations centered on the three thrusts outlined by the former Federal Coordinating Council for Science, Engineering, and Technology in its Advanced Manufacturing Technology initiative:

- "Intelligent" manufacturing equipment and systems;
- Integrated tools for product, process, and enterprise design; and
- Advanced manufacturing technology infrastructure.

Since these thrusts were outlined, the present administration has emphasized to a much greater degree the importance of a national information infrastructure that would support many activities of national importance, including manufacturing. That issue was considered by the committee as well.

The range and combination of research topics recommended by the committee are an essential feature of this report. Some of the topics chosen by the committee have been proposed by others in prior reports; the need for work in some areas is enduring. Some topics fall into areas of long-standing need but are advanced with new emphases. Because of limited time, the committee was unable to assess in depth the topics it identified. Moreover, it did not specifically address areas other than information technology that might prove beneficial to manufacturing; such areas include new physical processes that might be developed to shape and fabricate discrete components. In preparing this final report, the committee drew heavily on the preliminary report it issued in late 1993. Its efforts in subsequent meetings served to develop more fully, validate, and complement the ideas presented in that earlier report. A site visit to an engine plant in March 1994 (described in Appendix B) provided a firsthand demonstration of the productive use of current information technology in manufacturing. In addition, a National Science Foundation workshop in May 1994[3] (attended by some committee members) helped to crystallize some of the ideas presented in the material on the design process.

The CSTB and MSB are grateful to the National Science Foundation, to those who made presentations and/or submitted written material to the committee, and to those who reviewed this report and its predecessor. CSTB will be glad to receive comments on this report. Please send them via Internet e-mail to CSTB@NAS.EDU, or via regular mail to CSTB, National Research Council, 2101 Constitution Avenue NW, Washington, DC 20418. The committee, of course, remains responsible for the report's content.

---

[3] See Mukherjee, Amar, and Jack Hilibrand (eds.). 1994. *New Paradigms for Manufacturing.* NSF 94-123. National Science Foundation, Washington, D.C.

Contents

Information Technology
Technology
for
Manufacturing

Executive Summary

A manufacturing business is devoted to the production of tangible objects that are high in quality and competitive in cost, meet customers' expectations for performance, and are delivered in a timely manner. Finding and achieving the appropriate balance among these attributes—quality, cost, performance, and time to market—challenge all manufacturing businesses. Those companies that are successful in meeting that challenge remain in business; those that are not usually disappear.

In a manufacturing environment that is perhaps changing more rapidly now than during the Industrial Revolution, competing successfully will require that U.S. manufacturers increasingly provide customers with shorter times between order and delivery and between product conceptualization and realization, greater product customization, and higher product quality and performance, while meeting more stringent environmental constraints. Accomplishing these goals will require major changes in current manufacturing practices; such changes include the use of new and/or more complex manufacturing processes, greater use of information to reduce waste and defects, and more flexible manufacturing styles.

This report outlines a broad research agenda for applying information technology (IT)[1] to improving the manner in which discrete manufacturing processes will be carried out in the 21st century. These processes include the design of

---

[1] IT includes the hardware that computes and communicates; the software that provides data, knowledge, and information while at the same time controlling the hardware; and the interfaces between computers and the tools and machines on the manufacturing shop floor.

1

products and processes (e.g., converting customer requirements and expectations into engineering specifications, converting specifications into subsystems), production (e.g., moving materials, converting or transforming material properties or shapes, assembling systems or subsystems, verifying process results), and manufacturing-related business practices (e.g., converting a customer order into a list of required parts, cost accounting, and documenting of all procedures). This report also discusses the need for non-technology research to better understand human resource and other non-technical aspects of manufacturing.

THE POTENTIAL OF INFORMATION TECHNOLOGY IN MANUFACTURING

An enormous amount of information is generated and used during the design, manufacture, and use of a product to satisfy customer needs and to meet environmental requirements. Thus it is reasonable to suppose that the use of information technology can enable substantial improvements in the operation, organization, and effectiveness of information-intensive manufacturing processes and activities, largely by facilitating their integration (Figure ES.1). Equipment and stations within factories, entire manufacturing enterprises, and networks of suppliers, partners, and customers located throughout the world can be more effectively connected and integrated through the use of information technology.

Information technology can provide the tools to help enterprises achieve goals widely regarded as critical to the future of manufacturing, including:

- Rapid shifts in production from one product to another;
- Faster implementation of new concepts in products;
- Faster delivery of products to customers;
- More intimate and detailed interactions with customers;
- Fuller utilization of capital and human resources;
- Streamlining of operations to focus on essential business needs; and
- Elimination of unnecessary, redundant, or wasteful activities.

The development and implementation of new information technology to meet these goals will be shaped by organizational, managerial, and human resource concerns that have prevented manufacturers from exploiting fully even the technology that exists today. Sensitivity to these concerns will be essential to the successful development and implementation of the information technology associated with visions of manufacturing for the 21st century.

Information technology can be used to meet a range of needs of manufacturing decision makers (Box ES.1). These needs suggest a research agenda with both technological and non-technological dimensions; the primary targets of this research agenda include:

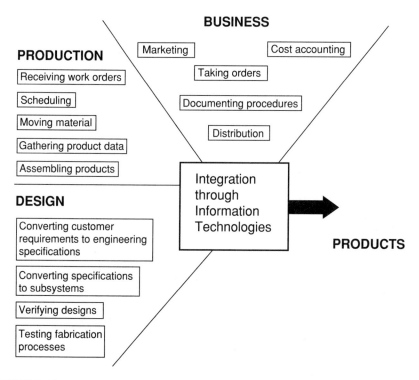

FIGURE ES.1 Information technology as a means to integrate various basic manufacturing activities.

- Operational control of factories and their suppliers,
- Tools for product and process design,
- Modeling and simulation of the entire spectrum of factory operations (virtual manufacturing), and
- Enterprise integration and use of other capabilities provided by the evolving National Information Infrastructure to support 21st-century manufacturing.

Other aspects of manufacturing, specifically physical processes, are not addressed in this report except as they relate to information technology's potential role in controlling them.

KEY FINDINGS

The key findings of this report are the following:

Finding 1: Information technology has a major role to play in the manufacturing environment of the future. The information-intensive nature of manufac-

BOX ES.1 Needs of Manufacturing Decision Makers and Examples of How Information Technology Could Contribute to Meeting Them

Need	Example of Information Technology's Contribution
Situational awareness. Both white-collar and blue-collar personnel must be informed about events in the manufacturing environment. An unexpected event may be anything from the breakage of a tool or the delay of a shipment to a design change made to a product.	To promote and enhance situational awareness, an IT-based factory information system could display the status of various tools and machines on the shop floor.
Diagnosis of problems. Decision makers need to identify the nature and extent of problems. Unexpected events can have a variety of causes. For example, a tool may cease functioning because it blew a fuse, because the bit broke, or because the motor seized due to a lack of lubrication. The stoppage could also have been the result of another error or problem somewhere else on the shop floor. Knowing what caused the problem is key to fixing it.	To assist in problem solving, diagnostics aboard a tool could be transmitted to a shop steward in real time.
Analytical tools. Decision makers need to evaluate and test various problem-solving approaches and strategies. For example, a decision maker may need to choose between allowing a cell to operate at reduced speed (lowering the throughput but also the risk of damage) or operating it at full speed (increasing the likelihood that the tool will have to be shut down entirely for repairs).	To enhance analytical capabilities, information technology-based simulations could help factory managers understand the consequences of different courses of action.
Dissemination channels. Solutions to problems must be disseminated. For example, information about the appropriate speed choice for the tool described above is needed both by the on-site crew and by the machine's manufacturer.	To enable timely dissemination of solutions to problems, information technology networks can be used to provide relevant text and graphics to all affected sites.

turing across all component activities will place ever-higher premiums on the faster, more accurate, and more useful presentation and analysis of the information streams associated with these activities. More, and more timely, information will be needed for decisions on the shop floor and in executive suites alike; information-intensive programmable processes and automated on-line control can improve product quality and manufacturing flexibility; greater integration among design, production, and marketing will lead to a greater need to share information. Only by judicious use of information technology will manufacturing personnel from chief executive officers to factory floor workers be able to assimilate and use these streams of information.

At the same time, information technology is only one dimension of enhanced future manufacturing operations; other important dimensions range from a better understanding of basic science and engineering phenomena in various domains to insights into the organizations and institutions that will be the users of advanced information technology. Progress in these other dimensions is also needed.

Finding 2: Current information technology is inadequate to support the manufacturing styles and practices that will be needed in the 21st century. Moreover, although individual demonstrations in particular manufacturing activities today hint at the potential impact of information technology, high degrees of integration have not yet been achieved. Open architectures and standards (currently absent) are needed to attain higher degrees of integration, because as the capabilities and applications of the National Information Infrastructure increase, the penalties associated with closed or proprietary manufacturing systems will only grow. Better information technology will also contribute to major improvements in product and process design and to more efficient and flexible shop floor operations, as well as to the planning and business capabilities of factory managers. Research on information technology in a manufacturing context is needed to enlarge the flexibility available in the future for manufacturing managers.

Finding 3: Exploiting the full potential of information technology to improve manufacturing will require addressing many non-technological matters, as well as the technical areas. For very good reasons, manufacturing managers are cautious about the promises of information technology and concerned about underestimating the potential risks of investing in it; these managers will require information technology that meets their needs as they understand those needs. Questions of technological risk, work force adaptation, and organizational resistance cannot be dismissed; if information technology developments are to be viable and acceptable in the factory environment, such questions must be directly addressed and resolved.

The potential benefits of improving manufacturing performance are enormous—they relate to the basic good health of the domestic and international economies. On the other hand, the risks are significant; the business landscape is littered with the hulks of companies for which the use of technology did not solve deeply rooted problems. This report seeks to identify research directions in information technology that can help to support and sustain a vibrant U.S. manufacturing enterprise.

RECOMMENDATIONS

The committee's recommendations for research fall into two broad categories: those related specifically to increasing the sophistication with which it is possible to apply information technology to manufacturing needs (a technology research agenda) and those related specifically to increasing the likelihood that such technologies will indeed be used appropriately in future manufacturing endeavors (a non-technology research agenda). Given the charge of the committee to focus on a technology research agenda, recommendations related to the first category are emphasized. In many respects, however, the non-technological areas require more attention, since such areas are usually the most difficult in which to effect change.

A Technology Research Agenda

The ultimate goal of the information technology research agenda outlined in this report is to expand the envelope of technological options and capabilities for managers of 21st-century manufacturing businesses. But it is reasonable to separate research efforts that can have an incremental impact on actual manufacturing operations in the nearer future from those that may have a revolutionary impact in the farther future. Research that is likely to have an impact in the shorter term is most likely to benefit from awareness of social and organizational issues.

With this perspective in mind, the research areas relevant to product and process design (Chapter 3) and shop floor production (Chapter 4) appear most promising in the nearer term. Factory modeling and simulation (Chapter 5) are farther-term, higher-payoff research areas. Nevertheless, it must be recognized that manufacturing is "an indivisible, monolithic activity, incredibly diverse and complex in its fine detail . . . [whose] many parts are inextricably interdependent and interconnected, so that no part may be safely separated from the rest and treated in isolation, without an adverse impact on the remainder and thus on the whole."[2] This fact largely precludes the identification of specific "silver bullets" out of which all other progress in the field would flow.

Integrated Product and Process Design

In the area of integrated product and process design, the following questions warrant attention:

---

[2] Harrington, Joseph. 1984. *Understanding the Manufacturing Process: Key to Successful CAD/ CAM Implementation*. M. Dekker, New York.

• *How should the information associated with products be captured and represented?* Issues relevant to this question include the representation of high-level functions for manufactured mechanical or electromechanical products, the creation of abstractions that contain the right amount of detail for their use at different points in the design process, formalisms for the representation of both domain-independent and domain-specific information, the interchangeability of product data models for use by different parts of the manufacturing operation (e.g., design, fabrication, test, maintenance, upgrade), and the relationships between high-level function abstractions and the physical reality of geometry and materials.

• *How can manufacturing processes be represented?* Process description involves languages for description and models of specific manufacturing processes, both as they actually exist and as they might be improved. Languages will have to express in compact form not only nominal process behavior but also variant behavior. They should have features that support checking for correctness and completeness and should be translatable across technical domains. Models of specific processes must include the information necessary to support dynamic control of individual operations and to take local environmental conditions into account, and they must faithfully represent real manufacturing processes as they exist.

• *How should tools be constructed that support product design?* An integral aspect of product design is how to make trade-offs (e.g., among cost, performance, and reliability; between alternate space allocations; between making or buying a component; between long-term operating costs and initial costs; and so on). Designers would benefit greatly from tools that would help them evaluate these trade-offs in a rigorous and systematic manner. Presentation and display tools for visualizing different design alternatives would also support the product designer. Finally, it will be essential to develop tools that support analysis of the "off-nominal" behaviors that result from manufacturing inaccuracies or deliberate variations introduced into a generic design.

Shop Floor Production

In the area of managing shop floor production, the following questions warrant attention:

• *How should shop floor tools be controlled?* Machine controllers are the fundamental interface between a factory automation system and the fabrication or assembly tools themselves. Research is needed that will result in an open architecture for machine controllers and in a good language for describing unit processing operations.

• *How should shop floor operations be scheduled?* Effective real-time dynamic scheduling of shop floor operations is central to ensuring the efficient use of shop floor resources. Effective real-time scheduling requires continuous tracking of the status of jobs, work cells, tooling, and resources and should support reactive scheduling and control (e.g., rerouting work flows to compensate for a problem or exploiting fortuitous windows of opportunity in individual work cells). In addition, real-time schedulers must give human decision makers convenient and transparent access to relevant information and tools enabling them to make trade-off decisions regarding release, reordering, sequencing and batching, and other matters. Finally, new production planning and scheduling optimization techniques are needed.

• *How should sensors be integrated into the shop floor environment?* Sensors provide accurate real-time information on what is happening on the shop floor and at individual work cells that is not available through other means. Such information is needed to guide unit processes as well as to provide status information and situational awareness at higher levels of authority (whether automated or human). The future manufacturing facility will use many different types of sensors, creating a requirement for a standardized control system into which sensors can be plugged with minimal bother. Reliable operation will also be at a premium.

Factory Modeling and Simulation

In the area of factory modeling and simulation, the following questions warrant investigation:

• *How can an individual production line be simulated?* Although it will not be possible to fully simulate even a modest factory for many years, it may be possible to simulate individual production lines. Research in this area would build on the single-activity models already in use in manufacturing to integrate their functions and to provide a comprehensive overview of the production line.

• *How can virtual factory simulations be validated?* A key question in any simulation is the extent to which it provides an accurate representation of reality. Simulations can be tweaked and otherwise forced to fit empirical data, but since the purpose of a factory simulation is to make reliable predictions about a factory operation for which there are no empirical data, managers will need strong assurances that a simulation's prediction of a new factory configuration faithfully reflects what would actually occur. Research is needed to develop methods of validation that provide such confidence.

• *How is the complex and voluminous information associated with a factory best presented to users?* User perspectives and needs are not homogeneous, yet the information presented to different users will all be derived from the same data sources, whether a factory model is operating in control mode or simulation mode. The visibility and transparency of activities to all relevant users across the enterprise are prerequisite to achieving effective control.

• *How can the consistency and accuracy of concurrently used models be guaranteed?* A factory model is most likely to be some aggregation of smaller models, each built to represent or simulate or control some single factory activity. Ensuring that the assumptions underlying these models and the data streams driving them are consistent across the board will be a major challenge.

• *How should dynamic interactions among these interconnected and interrelated models be understood?* As smaller models are aggregated into a large factory model, it is inevitable that they will interact with each other (since the processes and products they represent also interact with each other). Understanding the nature and scope of these interactions will be a major challenge with important implications for ensuring model fidelity and validation.

Information Infrastructure to Support Enterprise Integration

Electronic networks and related elements of information infrastructure are likely to be the means for achieving a relatively complete integration of the manufacturing enterprise, including activities within a given firm as well as activities undertaken by suppliers and customers outside the firm.

The following questions suggest research areas relevant to enterprise integration:

• *What standards should support the passing of information between the various architectures and the interconnection of different systems within the manufacturing enterprise?* Today, incompatible representations of knowledge and information are common in computer-aided design, computer-augmented process planning, and computer-aided manufacturing. These incompatibilities are major obstacles to enterprise-wide integration.

• *How can standards be made to accommodate some upgrade capability?* In the absence of an upgrade capability, technology vendors worry about premature freezing of technology and customers worry about intrinsic obsolescence.

• *How should real-time communications architectures be implemented?* Robust real-time control systems for tools such as cutters will demand network architectures that can tolerate dropped messages or delayed message arrival. Thus research is needed to formulate the principles for construction and operation of networks that can support time-critical message delivery in a context of interconnecting, multipurpose networks.

• *What tools and capabilities are needed for human- and machine-based information and resource searching?* These essential capabilities should become part of the underlying network service infrastructure in order to increase network performance and efficiency.

A Non-technology Research Agenda

The issues relevant to exploiting the full power of advanced information technologies go beyond traditional engineering research. Researchers in economics, organizational studies, and management science may contribute a great deal to understanding how manufacturing enterprises can actually make use of such technology. Non-technical problems are often exacerbated, for example, by the globalization of industry, in which the relationships with suppliers, customers, design centers, and factories are increasingly distributed over a wide band of cultures, time zones, and expertise. Those wishing to accelerate the adoption of advanced information technology for manufacturing must ensure that human, organizational, and societal factors are aligned so as to support its acceptance and maximize its benefits. Many mechanisms can facilitate such alignment, including sabbatical programs for industrialists and academics in each other's domain, teaching factories created to prepare future manufacturing specialists, and advanced technology demonstrations that illustrate the benefits of information technology for factory performance.

In addition, considerable research in social science will be necessary to facilitate the large-scale introduction of information technology into manufacturing. New technologies generally require new social structures if those technologies are to be fully exploited. Innovators will have to confront issues such as the division of labor between human and computer actors, the extent and content of communications between those actors, and how best to organize teams of human and computer resources. Continual upgrading of skills and intellectual tools will also be necessary at all levels of the corporate hierarchy.

A particularly important step in supporting 21st-century manufacturing will be to develop accounting and financial schemes that enable manufacturers to account for their increasingly critical intellectual and information assets in the same way that the generally accepted accounting principles of today allow them to treat building and pieces of equipment—as capital expenditures (i.e., expenditures that relate to the long-term value of a company).

Finally, standards and metrics are a mix of both technical and non-technical issues. Standards are needed to support and facilitate interoperability and open architectures and systems, while metrics are needed to determine the impact of information technology on various dimensions of manufacturing. The technical work required in both areas is substantial, but the organizational and social issues that need resolving before appropriate standards and metrics are in common use also deserve much more attention than has been given to date.

1

Vision and Recommended
Areas of Research

INTRODUCTION

The manufacturing sector is the crucible in which many technologies are refined and fused for the purpose of making things that people need or want. In 1993 manufacturing accounted for 18 percent of the $6.4 trillion gross domestic product and for nearly 18 million jobs in the United States (U.S. Department of Commerce, 1994; Council of Economic Advisors, 1994). Broadly defined, manufacturing includes all of the activities involved in determining the needs of potential customers, conceiving and producing products to meet those needs, and marketing and delivering those products to the ultimate customer. Money is made and needs are satisfied by meeting quality, cost, performance, and time-to-market goals for the product being manufactured. These attributes—quality, cost, performance, and time to market—may be taken to be the yardsticks against which any new advance must be measured.

Given that U.S. leadership in certain areas of manufacturing is no longer the rule, it is reasonable to ask what needs to be done to regain international leadership in manufacturing. Suggestions abound, but in the absence of a clear strategy in this area, federal decision makers have struggled to find the right mix of investment in manufacturing research.

This quandary has extended beyond the funding agencies to the universities and to academic research. There are few departments of manufacturing at U.S. universities because many academics do not believe that manufacturing is an academic discipline. Moreover, the disciplines basic to making progress in manufacturing belong not only in the "hard" sciences and engineering in physics, mathematics, mechanical and electrical engineering, industrial engineering, computer science and engineering, chemistry, and materials science but also in the "softer" sciences of sociology, psychology, management science, and economics. Even though these separate disciplines are individually supported by funding agencies and universities, there is a lack of focused attention on how to integrate basic knowledge from many disciplines into knowledge that furthers manufacturing goals.

Information Technology and the Increasing
Complexity of Manufacturing

At the same time that this lack of strategy is apparent, all dimensions of manufacturing (e.g., products, markets, processes) are becoming more complex, diverse, and international. Indeed, common products such as automobiles can have thousands of parts, and modern aircraft and integrated circuits include millions of parts or active elements. Each of these products takes years to design, requiring the effort of hundreds or even thousands of people worldwide. Complex new products based on information content and their accompanying information-dominated design and manufacturing methods already require us to deal with entirely new scales of complexity.[1] Some products require such levels of precision, delicacy, or cleanliness that people can no longer make or assemble the parts; in some cases, they cannot even see them.

To realize these and other products, manufacturing firms must cope with design processes (e.g., converting customer requirements and expectations into engineering specifications, converting specifications into subsystems), production processes (e.g., moving materials, converting material properties or shapes, assembling products or subsystems, verifying process results), and business practices (e.g., converting a customer order into a list of required parts, cost accounting, and documentation of procedures). The illustration on the cover indicates the relationships among these various elements of manufacturing and the role of information technology (IT; Box 1.1) in integrating them (see also Figure 1.1). By providing ways to facilitate and manage the complexity of these information-intensive processes, as well as to achieve integration of manufacturing activities within and among manufacturing enterprises, information technology will play an increasingly indispensable role in supporting and even enabling the complex

---

[1] A case in point is very large scale integrated (VLSI) chips. A single VLSI chip may have several million transistors with submicron feature sizes. A complex system may have hundreds of chips and tens of millions of transistors. Logic design, functional tests, fault tests, timing, placement, and wiring data run to gigabytes per chip. Validation of a design may involve many millions of simulated test cases. Finally, different aspects of chip design are coupled, so that changes required in the logic design (for example) often affect the analysis of derived fault, timing, and place and wire views of the logic. Similar observations apply to airplanes, ships, and cars.

BOX 1.1 Information Technology for Manufacturing— Definition and Elements

Although there are many definitions of information technology (IT), this report defines IT as encompassing a wide range of computer and communications technologies. IT includes the hardware that computes and communicates; the software that provides data, knowledge, and information while at the same time controlling the hardware; and the robots, machinery, sensors, and actuators or effectors that serve as the interface between computers and the outside world, specifically the manufacturing shop floor. Note also that the effective use of information technologies demands considerable investment in training and maintenance. Examples of IT include the following:

- Computers
    - Workstation
    - Mainframe
    - Server
    - Personal digital assistant
- Communications devices and infrastructure
    - Telephone
    - Local area network
    - Wide area network
    - Wireless network
- Software
    - Operating system
    - Artificial intelligence expert system for product configuration
    - Computer-assisted design package
    - Animation and simulation software
    - Virtual reality simulations
    - Software for total quality management and inventory control
    - Scheduling package
- Sensors
    - Machine vision
    - Tactile and force sensors
    - Temperature sensors
    - Pressure sensors
- Actuators or effectors
    - Robot arm
    - Automated ground vehicle
    - Numerically controlled cutter
    - Microactuators

Information technologies are focusing to an increasing degree on knowledge and information rather than data alone. That is, advances in information technologies over the last 40 years have enabled the manipulation and processing of increasingly abstract and higher-level forms of information. For example, industries cannot rely only on postmortem quality control data to detect product defects: modern quality assurance requires that potential problems be traced back through the manufacturing system for high-level analysis at each manufacturing unit. IT that is used in support of such an approach depends as much on knowledge and diagnosis as on simple data gathering.

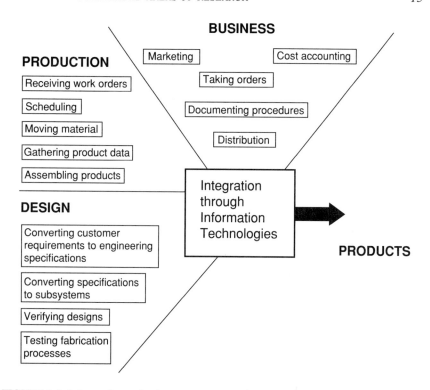

FIGURE 1.1 Information technology as a means to integrate various basic manufacturing activities.

practice of manufacturing. In the decades to come, information technology may have an impact on manufacturing performance and productivity comparable to that of mass production.

Purpose, Scope, and Content of This Report

This study was conducted to identify areas of information technology-related research needed to support future manufacturing. The committee chose to define manufacturing broadly as the entire product realization process, from specification through design and production to marketing and distribution. Although it believes that information technology has important applications to both continuous and discrete manufacturing, the committee focused on discrete manufacturing as the type in which the problems of applying information technology are most pressing. It did not include in its deliberations such important dimensions of manufacturing as the study of physical processes in manufacturing, although it did address information technology as it might be applied to controlling these processes.

Chapter 1 of this report outlines some of the technical and other challenges confronting the manufacturing enterprise at the outset of the 21st century, expresses a vision of future manufacturing based on what is known today and what might be expected from information technology-related R&D efforts in the future, and recommends a research agenda aimed at achieving this vision. Chapters 2 through 7 elaborate on the contents of Chapter 1. Chapter 2 presents the context for manufacturing. The R&D agenda implied by the vision of Chapter 1 is the subject of Chapters 3 (product and process design), 4 (shop floor production), 5 (factory modeling and simulation), 6 (information infrastructure issues), and 7 (nontechnology issues). Chapters 3 through 7 explore in more detail the research questions that must be answered successfully if the vision of a robust and internationally competitive 21st-century manufacturing enterprise is to be achieved. A list of contributors to the report, a description of an engine plant visited by the committee, and sketches of possible advanced long-range technology demonstrations are given in Appendixes A, B, and C, respectively.

FLEXIBILITY FOR THE FUTURE

In the manufacturing environment of the 21st century, several trends will place increasing pressure on manufacturers:

• Larger numbers of product variants will be required to meet user demands for greater product customization. This will lead to pressures to reduce production lot sizes while maintaining unit costs at an economic level. Manufacturers will need production facilities that are economic and profitable at very low volumes and that have low "fixed" costs.

• Increasing dispersion of manufacturing operations is likely. Successful manufacturing companies will be forced to develop effective global manufacturing networks, "knitting together" the skills and capabilities of individual units located around the globe to create a seamless international production capability; organizationally, the factory floor will see self-directed work teams "managing" the day-to-day operations of the firm with minimal real-time supervision, with white-collar labor focusing much more on the creation of new and improved products. These trends will almost certainly lead to a reduction in the average size of manufacturing facilities, the increasing use of "temporary" assets (via leasing or other cost-sharing arrangements), and the development of common processes so that manufacturing activities can be moved rapidly to locations that need increased production.

• Shorter time to market will become even more important than it is today. One aspect of this issue will be the ability to deploy technology rapidly. Another will be the ability to execute customer orders rapidly. Manufacturing concerns will emphasize work force skills and empowerment in order to meet marketplace needs for speed.

• Global environmental concerns seem certain to motivate the development of "green" and recyclable products and manufacturing systems that use fewer material resources.[2] Manufacturing systems should be able to accept "used" products that they have fabricated, disassemble the received items into component parts, and determine which parts are reusable or which are convertible into recyclable raw material with the ultimate goal of a product that leads to zero scrap material. Of course, products will need to be designed with such goals in mind.

Responding to these challenges will require unprecedented flexibility. Flexibility in manufacturing is associated with rapid responses at the appropriate level to new information and constraints, which may range from changes in consumer preferences or international trade regulations or union requirements to a temporary fault in a crucial piece of machinery on the factory floor. Whatever the source of change or constraint, information systems in the factory must enable an appropriate response. Information technology will enable better planning and organization as well, helping to control the events to which responses are needed.

Recognizing Information Technology's Increasing Capability in a Changing World

The role of information technology in manufacturing can be seen in the increasing use of computers to underpin product design and fabrication processes and to support related business processes such as sales and distribution. To date, the primary uses of information technology in manufacturing have been to control machinery and tools on the shop floor, to assist with administration in areas such as accounting and bookkeeping, to speed the transfer of information, and to support the management of product and process complexity (e.g., through computer-aided design (CAD) or manufacturing resources planning). Table 1.1 compares past and present characteristics and roles of IT in manufacturing.

Although these roles will continue to be important, the committee believes that information technology will become an increasingly significant source of support for different types of decision making needed in manufacturing (Box 1.2). This belief in the benefits of information technology is based on three premises:

---

[2] A recent agreement between two European auto manufacturers establishes a recycling network aimed at decreasing waste. It has begun to increase awareness of the importance of creating environmentally correct products for sustainable development around the world. In other cases, the producers of tires and batteries have been challenged to produce a product that can be disposed of without landfills. Taxes have been imposed on the purchase of new batteries and tires to help offset the cost of disposal. Yet such efforts are still the exception given the scope of manufacturing worldwide.

TABLE 1.1 Past and Present Information Technologies and Their Roles in Devices and Activities Integral to Manufacturing

	Past	Present
Information Technology		
Computing platform	Mainframe computers	Personal computers and workstations for most computing
Databases	Mostly paper records, stored in file cabinets	Large amounts of business data resident on electronically searchable, remotely accessible databases
Information retrieval	Human information specialists (e.g., public and private librarians and corporate information expediters)	Database retrieval systems now the basis for managing complex problems involving more subassemblies and more interactions with suppliers
Data communication	300 bits per second; hence, major restrictions on the size, complexity, and usefulness of the items communicated	1 to 10 megabits per second, often carried over local or wide area networks. Hence, large models (e.g., aircraft or automobile bodies) sent quickly, permitting designers states or continents apart to collaborate more easily
Manufacturing Technology		
Sensors	Mostly analog	Heavily digital
Recording media	Chart recorders	Computer-readable media
Control logic and machine controller	Mostly classical control theory (as exemplified by the proportional-integral-derivative controller); machine controllers using many subsystems based on programmable logic controllers	Classical control theory still used; modern control theory (state space analysis), fuzzy logic, and neural network controllers more common

TABLE 1.1 Continued

	Past	Present
Control systems	Analog amplifiers, electromechanical relays, and pneumatic and hydraulic actuators; automated machining operations performed using APT programming that generated cutter location data (CLData) on paper tape	Personal computers. Paper tape largely replaced by computer-controlled machines operating on CLData; data sometimes received through a computer network. Programmable technology enabling faster and more accurate control, with the end result that much more complex parts can be made much more quickly
Engineering Practice		
Analysis	Relatively minimal; manual processes; based largely on past practice—knowledge of what did and did not work in the past	Extensive and computer-supported to a large degree (computer-aided design to represent solid geometry of parts and assemblies, kinematic motions of parts, and some of the machinery used to make and assemble them); computer-aided engineering models of mechanical parts and assemblies used to simulate kinematic analysis
Design for product variety	Products largely standard, with few options for buyers	Higher degree of variety and customization possible
Product testing	Exhaustive testing of physical models	Computerized simulation and engineering analysis as substitutes for much physical testing; physical testing now used primarily as a final verification of design
Engineering style	"Over-the-wall" engineering, with market research, product design, and production working in isolation	Concurrent engineering (working on requirements, design, and production simultaneously) increasingly recognized as important and slowly becoming a common organizational objective, although not by any means the norm today
Problem resolution	Long face-to-face meetings between participants to address problems requiring attention by more than one department in a company	Electronic conferences, either by telephone alone or with video support, to address problems; meetings and meeting overhead thus reduced

continues

TABLE 1.1 Continued

	Past	Present
Production Operations		
Product data recording and use	Almost entirely paper records and engineering drawings	Electronic form for many types (especially for product or component shape and geometry)
Scheduling philosophy	Scheduling to maximize the use of (expensive) machines and people and to maximize work in process (WIP)	Scheduling to balance high levels of machine and personnel use and low levels of inventory (minimizing WIP)
Contingency management	Scheduling software task-oriented and not responsive to contingencies: software told what was to be done, and the human operator was expected to carry out the task	Software giving many shop floor personnel access to constantly updated information on status of machines, location of breakdowns, and schedule realization
Relationship between product engineering and product release	Systems completely separate and functioned nonconcurrently	Systems still separate, but operating concurrently
Information flow	Paper traveled with the product through assembly lines	In many factories, bar-code identification of parts moving through production that keep track of their positions and tell machinery which steps to perform; bar codes, coupled with digital status keeping, often used to develop systems that minimize "guess work" on inventory levels and improve use of assets

- Information technology will facilitate appropriate reuse of knowledge (e.g., reusing the design of a previously produced part rather than designing a new one from scratch), thus enabling decision makers to build on precedents and past decisions that have subsequently been validated by experience.
- Information technology will enable a high degree of integration among the various processes of manufacturing: product design and process design, shop

BOX 1.2 Needs of Manufacturing Decision Makers and Examples of How Information Technology Could Contribute to Meeting Them

Need	Example of Information Technology's Contribution
Situational awareness. Both white-collar and blue-collar personnel must be informed about events in the manufacturing environment. An unexpected event may be anything from the breakage of a tool or the delay of a shipment to a design change made to a product.	To promote and enhance situational awareness, an IT-based factory information system could display the status of various tools and machines on the shop floor.
Diagnosis of problems. Decision makers need to identify the nature and extent of problems. Unexpected events can have a variety of causes. For example, a tool may cease functioning because it blew a fuse, because the bit broke, or because the motor seized due to a lack of lubrication. The stoppage could also have been the result of another error or problem somewhere else on the shop floor. Knowing what caused the problem is key to fixing it.	To assist in problem solving, diagnostics aboard a tool could be transmitted to a shop steward in real time.
Analytical tools. Decision makers need to evaluate and test various problem-solving approaches and strategies. For example, a decision maker may need to choose between allowing a cell to operate at reduced speed (lowering the throughput but also the risk of damage) or operating it at full speed (increasing the likelihood that the tool will have to be shut down entirely for repairs).	To enhance analytical capabilities, information technology-based simulations could help factory managers understand the consequences of different courses of action.
Dissemination channels. Solutions to problems must be disseminated. For example, information about the appropriate speed choice for the tool described above is needed both by the on-site crew and by the machine's manufacturer.	To enable timely dissemination of solutions to problems, information technology networks can be used to provide relevant text and graphics to all affected sites.

floor operations, and business practices, thereby increasing the ease with which a product can be brought to market and reducing costs.

   • Information technology will increase the opportunity for human decision makers to think about the products and processes of manufacturing in abstract, higher-level terms without focusing on lower-level and repetitive details, thereby increasing the speed with which decisions can be made and implemented and improving the quality of those decisions. This premise is perhaps the most controversial of the three.

   These premises have been articulated and tested before, with varying degrees of success. What is it that gives rise to the committee's belief that information technology will be the basis of the next paradigm shift in manufacturing and a source of enhanced productivity?

   The most straightforward answer is that the world has changed. The cost-performance relationship for information technology has improved so much over the last decade that it now seems feasible to devote many more computational resources to problem areas that were previously starved for such resources. More importantly, social and technological factors relevant to the successful implementation of IT in manufacturing are now much better understood. Concerted attention to these factors will dramatically improve the prospects for using IT successfully in manufacturing in the future.

   For example:

   • Socially, the culture of manufacturing is, for many good reasons, highly conservative, whereas IT is an enabler and facilitator of radical change (although such changes may take place over time). Moreover, although manufacturing is hundreds of years old, computers have been available for use in factories for only about 40 years and thus have not yet been fully integrated into the factory. (Social dimensions of the resistance to IT are addressed further in Chapter 7.)

   • Technologically, early ventures in applying IT to manufacturing reached too far too fast. In contrast to the Japanese approach that blended IT applications in manufacturing with existing work forces, the U.S. approach was capital-intensive and tended to downplay operations and maintenance issues. In addition, the success of IT in many individual aspects of manufacturing has not been reflected in the integration of these applications into a smoothly running system, and such integration has been (and continues to be) quite difficult. In the absence of integration, it is difficult or impossible for different computer manufacturing applications to exchange information, and the result may be as cumbersome as having no automation at all (Box 1.3 describes a not-atypical experience in today's factories).

BOX 1.3 Elements of a Nonintegrated Computing Environment for Manufacturing, Circa 1990

Application	Computing Environment
Computer numerically controlled machine tools	APT running on closed architectures
Robots	Machine-specific languages such as VALII or AML
Day-by-day scheduling of orders	IBM personal computers running DOS; applications programmed in BASIC
Production planning	More powerful Unix-based workstations; applications programmed in C++
	IBM personal computers with DOS and BASIC
	Symbolics machines and applications in LISP for sophisticated constraint-based reasoning
Computer-aided product design	Unix-based Sun, DEC, or HP workstations; applications programmed in C. IRIS Silicon Graphics machines for viewing of solid models. International Graphics Exchange Standard applications running on these systems are difficult to exchange.
Sales, marketing, and secretarial functions	Apple II-C computers
Payroll	IBM 3000 series running VM/CMS operating system and COBOL
Financial planning	IBM personal computers and Lotus 1-2-3

Result: each computer system is considerably different from its neighbors; each responds to local events but is not open and/or cannot communicate well outside its domain.

Balancing Current Needs and the Development of Future Capabilities

At the same time that it acknowledges IT's promise, the committee also recognizes very clearly that for good and proper reasons, managers of manufacturing enterprises are much more concerned about turning out products today than about improving their operations tomorrow. In particular, they regard new technology as beneficial only insofar as it can help them to achieve their tactical goals of meeting schedules and generating revenue for next quarter's balance sheet. Thus, it is not surprising that many manufacturing managers, especially those who have a great deal of practical experience with the day-to-day problems in manufacturing, make an argument that goes something like this: "We have real and immediate problems in U.S. manufacturing today that need to be solved if we are to compete successfully in the future. These immediate problems are the ones that should be given the highest priority for the near future, and research on advanced information technology whose payoff lies decades in the future, if ever, is a poor use of today's limited resources." In the absence of an explication of the detailed benefits of new technology, in particular information technology, and the impact of those technologies on the bottom line, a plant manager may quite reasonably adopt a "show me first BEFORE you think of modifying MY factory" attitude.[3]

The committee recognizes the existence of substantial tension between the conservative nature of manufacturing as an activity and the radical change implied by IT in the long term. Clearly, the immediate problems of manufacturing warrant attention; if U.S. manufacturers do not survive because they are unable to solve today's immediate problems, all of our investments in technology research will go for naught. Much of what is needed today must be delivered today by today's suppliers; research (i.e., long-term work) holds promise only for the longer term.

Yet some degree of investment for the longer-term future is warranted, and investments made today in IT research for manufacturing may have high payoff in the future. Indeed, the committee believes that pursuit of a coherent research agenda in this area would exploit U.S. strengths in computing and communica-

---

[3] Such concerns are buttressed by the history of the first wave of computer-integrated manufacturing (CIM). CIM, a concept advanced in the late 1960s for efficient factory production, encompassed a number of different visions of manufacturing under computer control and had both proponents and detractors in various segments of the manufacturing community. In its first decade or so, the strategic advantage of CIM was seen to be the considerable reduction of "blue-collar labor costs." Strategic planners hoped that U.S. products would then be made and assembled as cheaply as those in the newly industrialized countries where labor costs were relatively small. The capabilities and implications of the technology were overstated by the strategic planners of the time, and they did not recognize the sea change in the nature of business operations as corporations became more transnational and the idea of simply using cheap labor was already fading. In this context, both the benefits and failures of CIM were obscured.

In concrete terms, the outcome of the first wave of CIM was mostly the installation of computerized machinery and robot arms on the factory floor, often in inappropriate applications or without the necessary expertise to use these systems. Much of the initial CIM investment provided a poor return, and today true computer-integrated manufacturing is far from commonplace in U.S. factories (as described in Box 1.3).

For more discussion of CIM, see Merchant (1971), Bjorke (1979), and Harrington (1984). For reference to the increasing transnationality of manufacturing, see Bartlett and Ghoshal (1992).

tions and enable the United States to regain a role in setting the world standard in manufacturing. Further, since technology is relatively easy to diffuse over national borders (and competitive advantages due to the possession of a given technology thus tend to be transient), a continuing role for research on IT related to manufacturing should be anticipated by policymakers and manufacturing managers.

Resolving the tension between taking care of immediate needs and investing in research with longer-term payoffs will require managers to understand the strategic importance of new technology even as they are enmeshed in their tactical environment, and technologists to develop new technologies with business goals in mind.

Looking Ahead

The committee fully recognizes that IT by itself is not a panacea. For example, a study of the auto industry by the Massachusetts Institute of Technology found that highly automated auto plants achieved only average productivity, and even though the automation was capable of substantial flexiblity, these plants produced only two body styles of one product.[4] But even in the research domain, research to fill the gaps in the non-IT-related scientific and engineering knowledge about products and processes to be supported by information technology is essential if the promise of IT is to be exploited fully. For example, a deeper basic understanding about materials and fluid behavior may be needed to support new fabrication processes or to improve old ones.[5] A deeper basic understanding about fatigue and corrosion may be necessary to support product designers attempting to reduce maintenance requirements. A deeper basic understanding of relations between tolerances and function may be key to developing new assembly and shaping processes to be controlled by IT.

A critical concern for developers of technology should be identification of the decision maker who will decide whether or not to adopt a given information technology innovation. Such decision makers are found at nearly all levels in a company's hierarchy, but they have different concerns depending on where they sit, and they look to new technologies to answer different questions (see Chapter 7, Figure 7.1, which indicates the types of questions that may be asked at various levels of an organization's hierarchy, the relationships between the various levels

---

[4] See Dertouzos (1989). Similar examples in which the use of information technology does not correlate with marketplace success have been found in the service industries as well; see CSTB (1994a). The lesson is clear—companies (either manufacturing or service) that use information technology inappropriately are not likely to reap significant advantages from such technology.

[5] New processes may include laser processing, water-jet cutting, and deposition-based fabrication. Chemical vapor deposition is an example of a process that was introduced only recently but is now common.

of the hierarchy, and the need for consistency and coherency throughout the enterprise). While the identification of these basic concerns is not IT research per se, technology researchers who wish their innovations to be adopted must craft and present their research in ways that address the needs and business aspirations of the decision makers in manufacturing. Without solving problems relevant to chief executive officers, technology researchers will seldom see their creations adopted.

Finally, in considering IT's potential for contributing to improvements in manufacturing, time line is a concern. The committee's vision is based on projections of how what is known today in IT and its applications to manufacturing might plausibly develop in the future. But forecasting time lines is difficult, and committee members' views on this issue reflect a spectrum of opinion. Even under the most optimistic view, it is not really plausible that most manufacturers, especially smaller ones, will have achieved this vision by 2010. However, it may be that major parts of it will have been achieved by a number of large manufacturing concerns, and that smaller firms will be learning how to use new information technology (or at least that they will be learning how to conduct their business in an IT-rich environment). Under a more pessimistic view, large-scale adoption may not happen until much later.

The committee emphasizes that its admittedly expansive vision for IT in future manufacturing should be recognized as just that—a vision—rather than as a definite prediction for the future. It is a vision driven both by threat (in that the vast power of information-driven manufacturing is increasingly recognized in other countries) and opportunity (in that the capabilities of information technology are growing at a rapid rate and that IT itself represents an area of U.S. comparative advantage), and describes the areas that are most likely to be essential to achieving a competitive (or indeed a leading) position in information-driven manufacturing.

THE POTENTIAL IMPACT OF INFORMATION TECHNOLOGY ON THE MANUFACTURING ENTERPRISE

The Broad Vision

In a future manufacturing enterprise characterized by ubiquitous and integrated computing, IT will be important in every aspect of manufacturing. Computers will be everywhere—on factory floors, in products, in offices, in wholesale and retail outlets, in homes, and on the street. Computers will be embedded into products as invisibly as electric motors are today. Familiarity with personal computing and use of a national information infrastructure will be widespread, and the intimidation factor that currently often prevents the consideration of IT solutions will be greatly reduced. Manufacturing decision makers will use IT to make real-time determinations of, for example,

- How best to respond to customer demand (e.g., corporate officers);
- Which process flow path to follow and when (e.g., area supervisors);
- How to respond to out-of-control systems (e.g., area process engineers);
- What material requirements exist (e.g., equipment operators);
- What optimal inventory levels to maintain (e.g., corporate managers and factory managers);
- How production activities are performing (e.g., factory managers); and
- What the project and unit cost will be (e.g., corporate managers and factory managers).

In addition, the IT supporting diverse queries will be seamlessly integrated, so that information needed from one part of the enterprise by another part will be transported with minimal difficulty. For example, in the manufacturing enterprise envisioned by the committee, plans for new processes or products will be transferred electronically from development into production, significantly reducing the interval between the design and realization of a process or product. Simulation models at various levels of detail (from floor operations to strategic planning) will couple to each other, so that results of one simulation model can serve as input to another.

If this vision of IT-enabled manufacturing comes to pass, future manufacturing operations will realize many further and significant improvements in:

- Time to market, through the analysis and use of appropriate "what if" scenarios;
- Factory layout, through the use of virtual factory models;
- Capacity and asset utilization, through the use of intelligent schedulers and rapidly reconfigurable factories;
- Yield, through the use of adaptive process control;
- Times for product and process transfer, through the direct transfer of design information to the production process;
- Matching product features and capabilities to customer needs, through increased customization and feasibility of economic small-lot production;
- Hands-on training, through the use of realistic models;
- Equipment performance, through use of expert systems and artificial intelligence technology; and
- Reduction in inventory and working capital, through better scheduling and forecasting algorithms.

Improvements in the areas above through the use of IT would represent a fundamental paradigm shift from today's manufacturing enterprise, one that may already be under way (as suggested by Appendix B). The following sections elaborate the committee's vision of how various dimensions of manufacturing (product design, process design, production, and business processes) may be transformed by information technology; a further discussion of these dimensions themselves is found in Chapter 2.

Nearer-Term Prospects for Improvement

Product and Process Design

Product design has expanded in scope in recent years. It has been traditionally understood as the collection of geometrical, material, and system specifications that achieve functional performance as a finished product that meets customer perceptions of need; today it also includes attention to manufacturability, usability, and environmental concerns. Thus, a designer may well choose a product design with inferior performance characteristics in certain noncritical aspects that is much simpler to produce than the alternative. To a much greater extent than is true today, 21st-century design will address product design aspects that are not often associated with traditional design at all, such as designing so that servicing a product will be easy and error-free or so that a robust final product can be made from parts obtained from many different sources. Different dimensions of product design will thus need to be integrated to an unprecedented degree.

IT will continue to help to improve the quality of designs and reduce the cost and time needed to produce them. In particular, the 21st-century design environment envisioned by the committee would allow product designers to create a "virtual" product and make extensive computer simulations of its behavior without supplying all of its details, and then "show" it to the customer for rapid feedback. The ability to undertake rapid electronic prototyping of designs would increase the ease of revising them, resulting in fewer design compromises and lower costs for making those changes. Product data would be represented in a uniform manner and would include information on all variants. Simulations would also enable the exploration of all of a product's behavior modes, including nominal behavior and major off-nominal variant behaviors that might affect performance, product quality, environmental quality, manufacturability, or user safety. The product design environment would provide computer assistance that would build on existing design knowledge and even existing designs but that would also be flexible enough to accommodate design innovation.

The designer would be able to specify a product in terms of function and performance rather than in terms associated directly with the production process (e.g., shape, tolerances, electrical inputs). For example, the designer of a motor would be able to use function-relevant parameters, such as the load capacity and life of a bearing or the allowed vibration frequencies and minimum fatigue resistance of a shaft. In addition, he or she would be able to express the desired functions of the product in a translatable and analyzable format that would permit decomposing a given design into functions and subfunctions and mapping these into engineering assemblies and subassemblies.

Whereas today's product designer builds a physical prototype or uses physical construction aids in the design process, tomorrow's product designer will certainly make extensive use of computer-based models. The construction of a physical prototype for use in testing a function or developing a tool necessarily freezes the design at the moment of construction and often leads to a loss of synchronization between model and product design when the design is altered after the prototype is built. An example of a more flexible approach to design is provided by the Boeing 777 airliner, whose computer-based design was undertaken by Boeing using a three-dimensional representation of a solid model. This application eliminated most paper documents and dramatically reduced design errors as measured by changes and corrections required to engineering drawings. Engineers were able to test small parts and whole sections of the fuselage to determine, long before these components were built, whether they would fit together properly on the factory floor. The capability for testing clearances between subassemblies to ensure human access was especially useful. Once conceptualized, changes to the virtual 777 could be realized immediately and communicated to all individuals interacting with the model, including distant contractors responsible for building components and subassemblies.

Another advantage is that these computer-based models often can enable different aspects of a product to be developed simultaneously. In the design of programmable electronic systems, for example, a linked product-and-process design environment would allow the supporting software to be tested on hardware simulators before the completion of hardware development. Since software development is often at least as time-intensive as hardware development, the use of simulators could cut overall development time by as much as a factor of two.

Extensive computational support would enable designers of manufacturing processes to explore, trace, and compare the production cost, quality, performance, safety, maintenance, and environmental implications of design decisions. The process designer would be able to specify a manufacturing process unambiguously and in ways that would yield information about, for example, its efficiency in advance of its actual deployment. Information technology would enable the process designer to identify the right manufacturing, assembly, and test processes for creating and verifying the elements of a product and matching them to specifications for functional behavior. He or she would also be able to develop models that could predict the expected results of variations implemented during design of a process. Such activities, which might be undertaken concurrently, could result in fewer errors, fewer repairs, shorter times to market, and lower cost. For example, flows of material through a production facility might be improved by evaluating various factory equipment layout concepts—to reduce the distance a product must travel before its completion or the number of handoffs between process steps, or to allow people to perform more efficiently—or by evaluating different material-handling modes (e.g., whether or not to dedicate an automated guided vehicle to a particular process).

In computationally supported manufacturing, process design models would couple to production line implementations of those simulations. For example, code to control tools on the shop floor (e.g., numerically controlled machines, robots, and transfer systems) would be generated through simulations. If a product's specifications changed, the process model would be refined by the process designer, and such refinements would be reflected in the production line. When necessary, process designers would be able to modify manufacturing processes adaptively, taking advantage of the knowledge available at every step in a factory's entire manufacturing process in order to improve yield at a subsequent or preceding step. For example, detailed real-time knowledge of the status of each piece on the shop floor (both unfinished product and equipment) might enable managers to optimize equipment use and/or lot movement globally, rather than locally improving each step but possibly reducing overall factory performance.

With product design and process design coupled electronically, new processes or products would go directly from electronic blueprint or simulation to production; use of the process model as the blueprint for process transfer would significantly reduce process or product transfer times.

The common theme in all these design applications is that the manipulation of information (in the form of design simulations and the like) is likely to be much cheaper and faster than real experimentation within an operating manufacturing facility. As a result, the design space within which it is feasible to explore alternatives is much larger, giving designers more options for how to craft a product that will meet user needs.

Shop Floor Production

In the 21st century, order-driven production (Box 1.4) will increase compared to today's forecast-driven production (i.e., production volume sized in accordance with a forecast of what the demand for the product is likely to be), which runs a significant risk of overproduction (with the unsold products incurring carrying costs charged to the producer) or underproduction (in which case production must be ramped up at great expense on an emergency basis). Under a "build-to-order" strategy, production volume is much more closely correlated with orders, thus reducing the risk of production that does not meet demand and providing considerable flexibility in changing production priorities.[6]

---

[6] Of course, when a customer needs an order filled in less time than it takes to produce a single item, "build to order" in a literal sense is not realistic. Thus, in practice, both forecasts and orders will influence production. Nevertheless, a production process can be configured to increase flexibility so that, for example, generic items can be produced in reasonable quantity and then customized on demand, thus reducing the time between customer order and fulfillment.

BOX 1.4 Order-driven Production

Order-driven production is rare today, though not nonexistent. For example, copy shops that use photocopiers to reproduce sets of university course notes on demand can be regarded as a kind of order-driven production shop. With high-speed laser printers coupled to large storage devices, printing books on demand is possible; a few such operations exist today, and many more are expected in the future.

An example of research in this area is found in the textile industry (often cited as a low-technology industry), which is currently undertaking research in virtual reality (VR) with the intent of recovering some of its $25 billion loss yearly due to clothing inventory markdowns and liquidation. Specifically, the textile industry, in cooperation with the federal government, is supporting VR research that will enable customers to shop for clothing in a virtual environment in which they would see virtual clothes on virtual images of their own bodies and feel how the clothes would fit. When a customer made a choice based on this experience, the order would be sent to a factory that would make these clothes on demand and then send them to the customer. Vendors would benefit from a reduction in the financial losses associated with product markdown and liquidation, while the customer would be provided with more choices and a better fit of garment to body.

For more discussion of the textiles example, see NRC (1994).

Factories in the future will continue to use a variety of technologies and processes and varying levels of automation. However, with the help of IT, fabrication of unique items could be more convenient and less expensive, and products could be manufactured economically in smaller lot sizes than is characteristic today and with greater ease of production changeover (smaller setup times). Scheduling of people, machines, and plants would be based on all aspects of an enterprise's operations, taking into account sales and market projections as well as work in process and capital utilization to dynamically schedule operations for the best profit. Scheduling would optimize overall factory performance, rather than the performance of local area functions as it does today.

In the 21st-century production environment foreseen by the committee, the new devices and processes used would result in very little material waste, with shaping and assembling devices, for example, extending their reach to ever smaller spatial scales. Increasingly, the materials used will be synthetics, composites, and ceramics. If used, robot manipulators would operate on ever-smaller parts and assemblies. Information would be embedded in parts and products and read by material-handling, shaping, assembling, and processing equipment, further automating the flow of materials and work in process. Parts would be self-identifying not only in the production process but also throughout the life of the product. Controllers of material-handling equipment would be more tightly coupled with machine controllers and orchestrated by higher-level software in accordance with global plant goals. New shaping and assembly devices would be capable of high-precision fabrication at unprecedented levels of repeatability. Tool and equipment control strategies would involve all the control techniques—classical, modern, fuzzy logic, and neural network controllers—in an integrated approach to control.

Production engineering would rely heavily on a reusable base of acquired production knowledge. Manufacturing production lines would be designed using modular design and control components interconnected through well-defined interfaces. When necessary, production lines would be reconfigured to accommodate new generations or types of products, and certain changes to the production process would be implemented in a very short time (perhaps minutes or hours) as external circumstances changed.

To meet the need for replacing or upgrading equipment, the production line would accommodate the tooling and fixturing necessary for adding new parts. Programs for controlling equipment (e.g., for numerical control) and systems (e.g., for scheduling and execution) would be generated and modified to accommodate changes to product requirements, changes in the production process, and changes in management objectives (e.g., little or no inventory, rush orders).

The components that make up a production line would be structured in a manner that they could be "plugged and played" together in a modular fashion. "Plug and play" would characterize the physical connections for machines (power and communication), the functional connections for operations, and the tooling specific to products being produced. Equipment in a production line would be easily moved, replaced, modified (upgraded), and enhanced (adding machines), improving system flexibility and responsiveness. Functional control modules such as scheduling procedures and planning procedures would plug directly into the factory control system directing the operation of shop floor equipment. Similarly, the modules affecting a product (e.g., tools, fixtures, masks) would be extendable to accommodate new products. Changes to a production plan would be possible with minimal interruption.

The 21st-century computationally supported production line would be robust in the face of an uncertain mechanical environment. Detailed information on factory and equipment status, based on data from sensors and controllers located on the shop floor, would be visible to managers and shop floor personnel alike. Real-time control would enable dynamic rerouting and reconfiguration of work flows to bypass problem areas in the production facility with minimal disruption; in worst-case situations, a production facility would exhibit graceful rather than catastrophic degradation. The ability of automated shop floor tools to undertake self-diagnosis and self-correction of routine problems would improve reliability and reduce the need for people to tend such tools. The extensive simulation capabilities of detailed factory models would enable companies to design facilities that incurred minimal operational difficulties and to train new employees in the use, maintenance, and repair of complex equipment without risking the cessation of factory operations.

Finally, the time needed to fabricate and deliver products would be significantly improved. Customized products would be delivered to consumers much more rapidly than today because of reductions in the time needed to make and assemble the various components of the product and in the time spent in waiting for processing by various components.[7]

Business Practices

Information technology could enable a high degree of integration between future manufacturing enterprises and their customers and suppliers. Working closely with a manufacturer via tie-ins to the manufacturer's information systems, for example, suppliers would know when to deliver supplies and would have blanket authority to deliver them as necessary. A precedent exists in certain car manufacturing operations today: suppliers query the manufacturer's production intentions and deliver their goods to the manufacturer's assembly plant in line sequence order, that is, in the order that they will be used. Information about needed supplies is thus passed automatically, without the manufacturer having to place an order explicitly. Suppliers may even be paid only when the final product is shipped, rather than when the parts are delivered, thus providing further incentive for just-in-time delivery. This practice, known as "pay on production," is increasingly common in the auto industry, because it lowers supply chain carrying costs that have created an artificial focus on inventory and it increases team play and product quality.

In computerized manufacturing, a supplier's own fabrication processes would easily accommodate changes required by the manufacturer. A customer's highly customized order for even a very complicated and sophisticated part (for which there may be many alternative choices) could be entered directly by the customer into a manufacturer's system: the order would be scheduled, and the customer would receive immediate acknowledgment and a product delivery date.[8] A request for material replenishment could go directly and electronically from a manufacturer's factory floor to the supplier's system. With protections against

---

[7] A good example is the manufacture of automobiles. The final assembly of a car takes a couple of days. Building the various components takes several days to a week. But the time from order to delivery for most American cars is usually at least several weeks. However, certain Japanese auto manufacturers now use sales forecasting techniques so sophisticated that most vehicles produced in a given production batch have a buyer by the time they come off the assembly line. In addition, the mix of options (a major variable in both throughput and production time) is roughly standardized to help match production batches with customer orders and is kept relatively small to simplify assembly. The result is that the time from order to delivery has been reduced to well under a week for these firms.

[8] Some manufacturers have begun to implement or plan such systems, although with less customization of product than that envisioned in this report. See, for example, the discussion of Motorola's fusion program for producing pagers in Trobel and Johnson (1993).

fraud and theft in place, the supplier would deliver the material directly to the factory floor, where receiving activities are performed. In short, the entire enterprise would be integrated all along the supply chain, from design shops to truck fleets that deliver the finished products.

Hints of this vision of integrated enterprises are available today, as businesses transfer information among related business organizations through electronic design interchange protocols. Some design information can be shared through common CAD systems or through information expressed through the International Graphics Exchange Standard. Suppliers to some manufacturers have been required to send progress reports of supplies in shipment and transit electronically for many years. A variety of industry-specific standards for transferring data electronically have been promulgated in the last 20 years. An advanced national information infrastructure such as that proposed by the Clinton administration would provide a widely accessible vehicle for linking manufacturers, suppliers, and customers to achieve the kind of integration and information sharing envisioned.

Factory decision makers' use of modeling would enable them to inquire about orders, market information, production status, product design, human resources, and financial information from a single unified source. Advanced modeling and analysis tools would make possible the collection and definition of data, processes, and associated knowledge at the depth and breadth necessary to support the simulation of business processes; models would take into account the form, meaning, and content of relevant data elements and processes. Dynamic simulations of an entire business would be possible, enabling decision makers to evaluate a wide range of "what-if" scenarios for the purpose of strategic and tactical planning. These simulations would operate much more rapidly than real time (simulating several months of factory operations in a few minutes), and so designers would be able to test a large number of alternatives for improving factory performance.

New Manufacturing Styles

If the nearer-term research challenges posed in this report are met fully and successfully and the anticipated nearer-term advances implemented so as to enable the capabilities outlined here, the committee believes that the result will be substantial improvement for manufacturing in cost, quality, asset utilization, productivity, environmental control, and time-to-market performance metrics. Implementation of these capabilities would also have profound implications for the workplace that would have to be addressed in a socially responsible manner. The longer-term concepts of a virtual factory and a programmable or reconfigurable factory as outlined below would embody IT to such an extent that the very character of manufacturing would be fundamentally altered. As presented, they pose targets and goals for research, whether or not they represent what future factories will actually be like.

The Virtual Factory

When a single factory may cost over a billion dollars (as is the case in the semiconductor industry), it is evident that manufacturing decision makers need tools that support good decision making about their design, deployment, and operation. However, in the case of manufacturing models, there is usually no testbed but the factory itself; development of models of manufacturing operations is very likely to disrupt factory operations while the models are being developed and tested. Today, decision makers have found that the use of good computer models of manufacturing facilities can provide valuable information that might otherwise have required time-consuming and expensive physical experimentation. More sophisticated versions of these simulations—what might be called virtual factories—call for a distributed, integrated, computer-based composite model of a total manufacturing environment, incorporating all the tasks and resources necessary to accomplish the operation of designing, producing, and delivering a product.

With virtual factories capable of accurately simulating factory operations over time scales of months, managers would be able to explore many potential production configurations and schedules or different control and organizational schemes at significant savings of cost and time in order to determine how best to improve performance.[9] Since a factory model running in simulation mode would run thousands of times faster than real factory operations and would likely cost much less as well, managers would have a rapid, nondisruptive methodology for testing various manufacturing strategies. Improvements suggested by real operations could be tested without risk in the simulation. Simulations could also assist in training tool operators and floor managers, who would be able to use factory models in simulation mode much as pilots use simulators to gain experience in flying real airplanes, especially under stressful or unusual conditions.

Computer-based factory models might also be coupled to real factories in what could be called "control" mode. In control mode, the factory model would actually control and run the operation of the real factory through manipulation of the objects in the virtual factory. Operating procedures and scheduling protocols would be validated in the virtual factory and then applied in or transferred to the

---

[9] Factory operation model development and testing are very different from process model development and testing in the sense that the disruption of an entire factory can be catastrophic for a firm's productivity. It would of course be possible to build a new factory that would carry the burden of experimentation. But factories are capital-intensive, and few companies are in a position to risk large amounts of money on time-consuming, expensive, or risky experimentation.

real production facility. Control mode would enable the direct electronic transfer of modularized capabilities from computer simulation to production line.

Coupled to appropriate computer-based reasoning and decision-support tools, a virtual factory operating in control mode would be capable of a significant amount of self-diagnosis. Driven by data from the real factory, the virtual factory would be able to analyze the performance of the entire factory continuously to determine the potential for optimizing operations to reduce costs, reduce production time, improve quality, or reuse materials.[10] For example, the virtual factory would be able to use the data collected by a factory monitoring system, analyze potential and actual failures, and identify the cause of a problem. Such a system assumes the availability of a knowledge base for every piece of equipment in the factory that, given certain monitored data, can be used in conjunction with a diagnostic system and reasoning and decision-support tools to identify the source of a problem.

Whether information is derived from a model run in simulation mode or control mode, results from good models can be examined from various user perspectives, including those of factory managers, product planners, and process equipment operators, in order to provide solutions to various types of problems that manufacturing personnel with different job tasks might encounter. Modeling and simulation are likely to become a basic tool used at all levels within the manufacturing environment, from senior management to equipment operator.

The Programmable or Reconfigurable Factory

A programmable or reconfigurable factory is one in which most or all of the information necessary for producing a product is embodied in a knowledge base and the associated programs, such that the factory can take in the information and produce the product with minimal human intervention. In such a factory, capital would substitute for labor both on the shop floor (with robots or other automated equipment undertaking many assembly and fabrication processes) and in real-time factory management (with computer-based decision-making functions operating on intellectual rather than physical inputs). Such a factory would be capable of a significant degree of "lights-out" operation, that is, operation with limited human involvement. Programmable or reconfigurable factories would also enhance managerial capability to cope with accidents and malfunctions on the work floor as work flow could be rerouted to bypass problem areas.

---

[10] Simple versions of self-improving factories have been demonstrated in semiconductor and chemical plants, where neural networks and/or adaptive control have been used to monitor and adjust parameters to optimize operations. Although such plants differ in nature and character from those for assembling discrete products such as automobiles or airplanes, it may be possible to develop analogous improvements.

Programmable factories expand the concept of the flexible manufacturing system, which is intended to be a set of machines and controllers that work cooperatively to manufacture similar parts in a product family in relatively small lot sizes but at unit costs and with qualities characteristic of mass production. But thus far, most such economically successful systems have produced parts with rather small variations in style and have exhibited limited software, little sensor feedback, and, still, relatively large lot sizes.

It is clear that a programmable factory would require software capable of operating on the entire range of product and process information. It would also require generic materials that could be used for different products made in the factory, generic tools and fixtures, programmable machine tools, and generic sensors. While not everything could be fabricated from generic materials, it might be possible to substantially reduce the number of different types of materials and tools needed to produce a diversity of parts. For example, it has been estimated that in a particular factory, 2,000 tools that were being used could be reduced to fewer than 50. In another case, 500 different types of steel could be replaced by 50 different types of steel. In a third case, 10 different types of 8-bit microprocessors could be replaced by a single 16-bit microprocessor. If more generic materials and tools could be used in production, high-volume purchases of these items could drive down costs.

In a programmable or reconfigurable factory, different products would be produced by changing software. Different software instructions would direct tool and machine controllers to perform different operations and to deliver different items to different work cells in different sequences. Although a single factory almost certainly could not produce computers and cars on different days, a highly programmable factory, in conjunction with new and more flexible fabrication processes, could produce cars in one week and trucks in the next week; the Toyota Motor Company does this in Taiwan today. Even without new fabrication processes, a highly programmable factory might be able to produce computers one day and defense electronics the next. Such a factory would depend on production facilities configured in such a way that a new production operation could be set up relatively rapidly. The ultimate goal is a paradigm known as "sell one, make one," as it is known in Japan.

In certain restricted domains, low-lead-time "reprogramming" of a factory to produce different products is possible today. The manufacturing facilities for books, very large scale integrated chips, and petroleum products produce many different products (different book titles, different chips, different types of fuel and lubrication) simply on the basis of a software change. Of course, the term "software" in this context refers to the particular text or masks that differentiate one book or chip from another, but the principle is the same—only intellectual inputs or changes are necessary to produce a different product, while the physical facilities of production remain much the same. But for other products, changeover times remain significant.

A programmable factory is also necessary for the economic manufacture of highly customized products. An example from today is the "on-demand" production of soft-cover textbooks in which chapters can be selected by a teacher based on his or her individual teaching needs. With increased product customization, customers would be able to obtain catalog items with the features, characteristics, and aesthetics they desired, at prices they could afford (probably comparable to the prices for mass-produced goods). The material and manufacturing cost per unit of producing a product likely would not be significantly different for a production run of 100 units or of 100,000 units in a single batch. Smaller lot sizes would also have major benefits with respect to quality control. When defects in a production process are caught early (as is the case when small lot sizes are produced), the amount of rework is minimized and fixes to the production process can be implemented more rapidly.

Note that high degrees of customization create additional stresses on scheduling. Since customization requires only small quantities of specialized materials, "just-in-time" scheduling either works properly or fails by idling the production machinery; the option of building inventory as a hedge against missed delivery times simply does not exist, since maintaining excess inventory is then a matter of purchasing unnecessary components rather than purchasing components that will ultimately be used.

The Networked Factory

In concept, a networked factory is one in which suppliers (both internal and external) and customers are connected electronically to a manufacturer (e.g., on the National Information Infrastructure). Manufacturers have been tied to suppliers and customers by telephone, mail, telex, and fax for years; the primary advantage of electronically networked connections would be the speed with which information could be exchanged and processed, sometimes automatically by intelligent agents that could respond to certain routine requests.

An electronically networked factory (hereafter a networked factory) would demonstrate significantly reduced transaction times as information technology reduced the delays of paper-based information transfer; information technology would facilitate instantaneous acknowledgment, scheduling of deliveries, and guaranteed service times. Many of the factors contributing to delays in the design and production processes would be significantly reduced within a networked factory. Reducing delay would contribute to reducing the time to market for new or improved products. A particularly important improvement would be a reduction of the time it takes a production facility to initiate the first step needed to respond to an order, since it is this time that often dominates the overall time required to fill an order.

Enabled through the National Information Infrastructure, networked factories would increase the options available to product and process designers.

Today's designers are strongly constrained by the process capabilities of manufacturers—product designers do not design products that their factories cannot make, and process designers do not create processes that their factories cannot implement. Indeed, even approaches to design regarded today as sophisticated (e.g., design for manufacturability, design for assembly) are necessarily limited by preexisting production processes and facilities. When a single firm owns the means of production, such approaches make sense. But if constraints on ownership are relaxed (and process elements consequently can be linked on a regional, national, or even international basis), designers can be freed to focus primarily on the expressed needs of the customer without worrying about how best to use a single plant for which many costs have already been incurred. Designers using a networked factory would be able to "outsource" various production processes more easily and to coordinate their operation.

Of all the different modern concepts in manufacturing, the idea of a networked enterprise including a networked factory is perhaps the most widely accepted and adopted; in some circles, the term "agile manufacturing" is also used. Further, the evolving National Information Infrastructure is expected to facilitate networking of all sorts. Chapter 6 discusses this connection in greater detail.

Microfactories

A microfactory is a production facility whose output capacity can be scaled up by the replication of identical facilities. Since microfactories would not depend on economies of scale for economic viability, they would draw strongly on the technologies of programmable or reconfigurable factories as they attempted to produce small-scale output at unit costs comparable to or only slightly higher than those for mass-produced items. If microfactories prove to be feasible, a single, large, centralized manufacturing facility could be replaced by a large number of replicated, modular microfactories that could be geographically distributed and located close to customers.

For producing quantities of identical items, traditional factories oriented toward mass production will probably remain superior to microfactories, because anything that could be done to improve the production process in a microfactory could also be done in a traditional factory. On the other hand, today's mass-production factories are capital-intensive construction projects that are themselves custom-built. If a microfactory could itself be mass produced in quantities large enough to reduce the cost of an individual microfactory, it might be possible to amortize the cost of designing the microfactory over many such (identical) facilities. Even today, some steel micromills have drastically reduced the capital cost of steel production. In addition, microfactories might incur lower product transportation costs (as the result of placing microfactories near customers), lower inventory costs (as the result of production on demand), or lower labor costs (as the result of using locally available labor).

Irrespective of cost issues, however, microfactories might well provide advantages in other situations. For example, microfactories might provide a way for firms to insert local content into manufactured products, perhaps through local final assembly—a capability that could be desirable for political purposes (e.g., as a dimension of international trade relationships). A second example of a microfactory could be a mobile fabrication facility (e.g., a microfactory located on a large naval ship that produces replacement parts for the battle group with which it sails); in such a scenario, economic concerns might be secondary to the capability for a rapid response. A third example is that microfactories of a sort do exist today, although they make business sense for reasons other than lower unit production costs. Microbreweries for beer and street-corner copy shops are two examples of microfactories in which production costs are higher than those of larger facilities; nevertheless, such microfactories fill niches because they provide higher quality or greater convenience. The primary challenge remaining for microfactories is one of economics.

GETTING FROM HERE TO THERE— THE NEED FOR BALANCE AND A CONSIDERED APPROACH

The various new manufacturing capabilities described above are tantalizing and appeal to many current notions of the progress possible in manufacturing. But for this vision to be realized, it will be necessary first to balance the responsibilities of factory managers and manufacturing decision makers to turn out quality products at low cost in a timely manner today against the desirability of planning to secure the potentially large improvements offered by judicious and innovative use of current and future information technology.

Even if these tensions are resolved, however, the full implications of successfully implementing IT are not known, and neither the committee nor the manufacturing community at large has thought through the many possible effects. To illustrate, success in some of the areas discussed above raises the following questions:

• If products are available on a fully customized basis, what happens to service, repair, and maintenance? Technicians may be faced with an extraordinary learning task if they are to be competent at repairing thousands of customized variations, although such a task might be mitigated by electronic information carried aboard the product itself. In locations far removed from production facilities, cannibalization of one unit to obtain spare parts for another is a time-honored maintenance practice that may no longer be feasible. Even today, documentation of new products for maintenance and repair technicians is an enormous problem—how will documentation be provided for an even larger number of products? Will information technology provide solutions to this problem, or will new approaches to product design reduce the need for voluminous documentation?

• What are the limits (if any) to desires for novelty and customization? Customers may not all want entirely customized products. Consider, for example, automobiles whose control systems (e.g., steering wheels, clutches) are entirely different; such differences might even prove detrimental to public safety. Cost pressures may limit the variety that consumers are willing to purchase. Customers may also resist products that demand that they learn new habits and operating procedures. Finally, customers may not even know what they want with the precision needed to specify a customized product.

• If automation replaces multitudes of manufacturing workers, what becomes of the displaced workers? Will they become managers? Technicians? Who will retrain them? For what will they be retrained? How will manufacturing workers in the new regime respond to being directed by automated overseers?

• If manufacturing operations are dispersed geographically, what becomes of team and corporate loyalties that are often the result of physical proximity and informal day-to-day social contact through work? What will happen to geographically based brand-name and corporate loyalties?

• What degree of information automation is "right"? In the case of physical automation, trying to automate many factories entirely proved to be a poor choice; improvements in productivity were obtained at the expense of flexibility, and it turned out that flexibility was a much more important characteristic. This may also be so in information management; what degree of automated decision making is appropriate? This is a very subjective decision, differing for different industries.

• How will research results, such as new fabrication processes, be converted into economical and reliable factory equipment? The same question applies to the conversion of new design algorithms into easy-to-learn CAD software. The industries that supply these vital infrastructure elements are short of technical expertise, financially weak, and subject to huge fluctuations in demand.

• How will the sophisticated ideas outlined in the committee's vision of future manufacturing be transferred to small businesses and lower-tier suppliers? Large businesses depend crucially on the lower tiers, but many businesses in these lower tiers may not be able to compete successfully without new technologies or assistance in adopting these technologies.

These questions, and many others, are largely outside the scope of this report. But their identification, articulation, and eventual resolution are an integral part of moving toward a vision of IT-enhanced 21st-century manufacturing.

Whatever one's view, it is clear that a number of major technological and sociological barriers must be overcome before IT's potential to revolutionize manufacturing can be widely accepted and achieved. More rapid progress in overcoming these barriers will increase the likelihood of earlier acceptance and achievement. The nature of these challenges and the research needed to overcome them are the subjects of the next several chapters.

THE RESEARCH AGENDA

Technology Research

In formulating a research agenda, the committee was faced with the realization that manufacturing is fundamentally a complex activity, with many interactions among its various components. In the committee's view, this realization reflects the nature of manufacturing as "an indivisible, monolithic activity, incredibly diverse and complex in its fine detail . . . [whose] many parts are inextricably interdependent and interconnected, so that no part may be safely separated from the rest and treated in isolation, without an adverse impact on the remainder and thus on the whole" (Harrington, 1984). Thus, it is fruitless to seek the identification of specific "silver bullets" upon which all other progress in the field depends. That said, however, the committee has identified several general themes for technology research that would advance the capabilities of information technology to serve manufacturing needs; these themes include product and process design, shop floor control, modeling and simulation ("virtual factory") technology, and enterprise integration as it affects factory operations and business practices.

The sections below summarize a research agenda that is discussed in detail in Chapters 3 through 6. The largest part of the research recommended in this report is aimed at developing various IT-based tools to support advanced manufacturing.

Product and Process Design

Product design and process design depend heavily on human judgment. Research is needed both to develop information tools that can help human designers make good decisions in their design work and to increase understanding of the design process itself. Enabling the creation of better tools and facilitating the design process could, for example, make it easier to generate a requirements specification that meets customer needs or to design a product or a process "from scratch" and/or through the reuse of existing and validated designs. Although better tools and techniques are needed in all fields, product and process design for mechanical components and assemblies is especially important.

The committee believes that a research agenda for product design (especially for the design of mechanical products) should build to the extent possible on the lessons learned in the design of electronic products such as integrated circuit chips. For example, the design of electronics today is based on having, at each stage in the design process, design abstractions that contain only the detail relevant to that stage. Using these abstractions, the product designer can postpone decisions about details and focus on higher-order questions about function, leaving the detail to the subsequent stages in which lower-level issues can be resolved. Designers of electronic products also have an extensive set of predefined and prevalidated "parts" that can be used as building blocks in product design; altering the parameters of these predefined parts allows some customizing of the product being designed. Such capabilities also need to be made available to designers of mechanical products.

Although mechanical products differ qualitatively from electronic products (e.g., mechanical products are three-dimensional, and interactions among their components are more analytically intractable), making their mechanical design as easy as the design of electronic products today is a reasonable asymptotic goal to work toward. In addition, tools are needed that will enable the identification of trade-offs between cost and dimensions of performance such as reliability, power consumption, and speed; between cost and design choices; between alternate space allocations; or in functional decomposition, subassembly definition, three-dimensional geometric reasoning, and make-or-buy decisions.

In the domain of process design, tools for describing processes are critical for the design of individual products, the design and operation of factories, and the development of modeling and simulation technology. Formal descriptions are necessary if processes are to be represented in sufficient detail and with enough specificity to be adequately complete and unambiguous; such formalisms would allow designers to describe factory processes (involving both machines and people), design activities, and decision processes, among others. Languages for describing processes must facilitate checking for correctness and completeness and must be able to express variant as well as nominal process behavior.

New tools for describing and representing processes could also be used to enhance product design, so that by simulation and emulation the best process could be matched to the product design (and vice versa) for maximum economic advantage (or to satisfy whatever criteria—such as quality or time to delivery—are important for the particular case). Used in this way, simulation and emulation could facilitate "design for manufacturability" and "design for assembly," which should also encompass design for rapid testing and diagnosis, fast maintenance and repair, quality control, and material handling, as well as design for modeling more efficient factory processes and operations.

Shop Floor Control

By automating processes, extending the uses of sensors, and improving scheduling, information technology can play a vital role in improving the flow of material and the routine control functions of machine tools, robots, automated guided vehicles, and many other basic machines on the factory floor.

Research is needed to advance the level of process automation, including greater ease of interconnection of factory equipment and more automated responses to problems. The current thrust today in systems development is toward open systems that allow equipment from different manufacturers to be "mixed and matched" as needed. Achievement of an open equipment controller architecture that would enable all factory and shop floor components to share the same programming environment, communication facilities, and other computer resources would contribute to the interconnection of factory equipment (as well as to enterprise integration). Advanced manufacturing languages that would be more flexible than existing languages for programming unit processes (such as APT and Compac for machining) would support the operation of accessory devices in conjunction with a particular process and be more closely coupled to product data generated by CAD/CAM systems (to facilitate direct transfer of products from blueprint to production).

Research is needed on advanced sensor systems as well. Sensors provide real-time feedback about the operation of a process during manufacturing (e.g., unpredicted part-tool interactions). Historically, sensors have served only as production monitors. Increasingly, they are becoming active components of production systems, integral to either a process or a finished product. Standardized sensor architectures must be developed so that sensors and actuators can be plugged into a common control system with only minor, automatic reconfiguration. Sensors connected through such architectures would be linked directly into databases for dynamic updates usable by machine controllers. Standardized sensor architectures will require a uniform method of characterizing sensors and actuators suitable for automation, applicable to a wide range of devices. Data fusion techniques for correlating inputs from multiple sensors would help overcome the difficulties of sensing in a relatively "dirty" environment. Intelligent sensors would be able to process shop floor data to higher levels of abstraction to determine their significance to manufacturing decisions.

Effective real-time, dynamic scheduling of factory operations on the shop floor remains a major problem but has great potential for improving factory performance. Dynamic scheduling is desirable because management priorities for production must be balanced moment to moment against circumstances prevailing in a plant and in the manufacturer's supply chain (e.g., sudden changes in conditions generated by drifts in machine capability, material shortages, unplanned equipment downtime, delays in arrival of necessary components). Dynamic scheduling would determine what should be done next by any particular piece of equipment at any particular moment based on current conditions throughout the factory. Research is needed to develop real-time scheduling tools that would provide capabilities for integrating scheduling and control reactively (e.g., tools that would make use of adaptive scheduling techniques based on the severity of the contingency at hand and the time available to adjust the schedule and/or scheduling techniques that could exploit windows of opportunity occurring fortuitously).

Also needed are information tools and techniques to ensure graceful degradation of plant operations in the event of local problems. Factory managers and operations teams will require effective means to support overall situation assessment, both within a factory (e.g., knowing when certain machines are inoperable, knowing the location of various parts) and outside it (e.g., knowing that a delivery will be delayed). They will also require tools that integrate multiple dimensions in which managers must make decisions, including decisions about product release, reordering, sequencing and batching, safety stock and safety lead-time, use of overtime, and order promising.

Autonomous agent-based architectures are a potential alternative to top-down scheduling. Autonomous agents (implemented as software objects or collections of objects, perhaps represented by physical robotic agents) could be attractive for manufacturing applications in the areas of planning, monitoring, and control. Important research problems connected with agent-based architectures include the level of autonomy that agents in various locations should have and how collections of agents would maintain stability when given potentially contradictory goals. The deployment of agents might well be risky until these issues are addressed in detail.

Modeling and Simulation

To realize a virtual factory that can faithfully reflect the operation of a real one in all relevant dimensions, it will be necessary to represent real manufacturing operations at different levels of abstraction. All objects in a real factory, whether they are pieces of equipment, product lots, human resources, process descriptions, data and information packets, or facilities, must have direct counterparts in the virtual factory; indeed, the actual production facility in which raw materials are transformed into physical products is itself one level of abstraction in a comprehensive virtual factory model. The boundaries of the virtual model must be flexible, capable of incorporating activities outside the factory or focusing only on entities within the factory structure as necessary for analytical purposes.

Central to simulation technology is research on modeling frameworks that can link the wide variety of models representing activities from all parts of manufacturing, from design through orders to multicountry manufacturing and distribution to customer delivery; a large number of models will be needed to simulate realistically even a modest factory. The models will be distributed in time and space; research will be required to understand how to link these essential pieces in a timely manner. Work is needed not only on general modeling techniques but also on fast methods of tailoring a specific model to local conditions. Hierarchical simulation models built from well-tested fundamental equipment blocks will be critical to the success of factory modeling. Similarly, procedures are needed that allow for the parallel processing of these hierarchical models so that very rapid (faster than real-time) simulation times can be achieved.

A second dimension of simulation technology is how to account for the stochastic nature of events on the factory floor. Given the multitude of unexpected events that can affect factory operations (e.g., tool breakdown, late supply shipments, personnel absences) as well as decisions influencing operations (e.g., which machine is to be changed, what personnel are to be used), no single simulation run will be definitive. Rather, tools are needed that can test a given configuration or plan in thousands of probabilistically determined runs.

A complete simulation that could be used for everything from analysis to control for even a modest factory is out of reach today. However, a first step would be the comprehensive simulation of an individual production line. For such a task, appropriately detailed models of individual tools are needed that can then be combined to provide overall realism. Equipment-level simulation models have been developed and used to analyze equipment-level characteristics, but the simulation of a production line would test the ability of such models to act in concert.

Validation of simulation models will be essential. Since a simulation is good only to the extent that it provides an accurate representation of reality, justified and well-grounded confidence in the model is critical for use and implementation. Simulation models can always be tweaked and otherwise forced to fit empirical data, but the purpose of simulation is to learn something reliable about a hypothetical factory operation for which no empirical data exist. Managers and decision makers will need high levels of assurance that a simulation's prediction of a new factory faithfully reflects what would actually occur, even taking into account the random events that affect today's manufacturing systems so adversely. Well-validated simulations would enable the creation of a demonstration platform that could compare results of a real factory system before the system ever operates. Tools to automate the process of sensitivity analysis for simulations would be particularly helpful in coping with the stochastic factory environment.

Ultimately, the modeling and simulation capabilities resulting from the research outlined here should be able to support configuring and constructing a real factory for high-level performance (on multiple dimensions), as well as planning how best to operate it once it has been constructed. A concrete demonstration of these capabilities would be the creation of a platform capable of comparing the results of real factory operations with the results of simulated factory operations using information technology applications such as those discussed in this report.

For modeling and simulation to serve manufacturing needs, two broad areas of research stand out for special attention: the development of information technology to handle simulation models in a useful and timely manner, and capture of the manufacturing knowledge that must be reflected in the models.

Enterprise Integration and Business Practices

Research is needed to extend and enhance the information infrastructure supporting manufacturing enterprises, including both the internal infrastructure used within a factory (perhaps a dispersed one) and the external infrastructure that increasingly links an enterprise to its suppliers, partners, and customers. The use of networks by all kinds of personnel and of equipment to exchange all kinds of data (text, numeric, graphic, and video) calls for high bandwidth; greater dependability and security; greater support for real-time communication, monitoring, and control; and better interoperability (through architectures, standards, and interfaces) for component systems and networks of different types. Achieving ease of interconnection is essential; attaching equipment and subsystems to a factory information system should be as easy as plugging household appliances into outlets, at least in principle. Beyond better network-related facilities, there is a need for better technology for the exchange of information, information services to support integration of applications, and standard representations, protocols, libraries, and query languages.

In addition, enterprise integration requires research and development relating to the interconnection of applications. Indeed, much of today's manufacturing information technology can be characterized as islands of automation that are unable to communicate with each other due to incompatibilities in their representation of largely similar information. Enabling intercommunication will require the development of appropriate ways of explicitly representing information related to products, fabrication processes, and business processes, as well as how each element relates to itself and to other elements. These new representation schemes will themselves demand a deep understanding of the underlying information, an understanding that is sorely lacking in many of the domains that relate to manufacturing.

Research is also needed on organizing principles and architectures for connecting different network-based applications into a seamless environment. Such connection is necessary, for example, to link flexible manufacturing cells to the plant scheduling function and to link the scheduling function to the enterprise order, delivery, and financial systems. Enterprise integration also implies a need for research to enable the automatic interpretation of the type of transaction being executed, the routing of the message to the right location for processing, and the processing that must occur when the message for the transaction reaches the correct system. Integration of individual enterprises into the marketplace in the information age will require security and authentication features that guarantee the integrity of electronic transactions.

Non-Technology Issues

Expanding the scope of what is achievable by information technology is only one dimension of realizing a 21st-century vision of manufacturing. It is equally important to understand how manufacturing enterprises can actually make use of the technology. Even today, much useful technology remains unused. Innovators in manufacturing must ensure that human, institutional, and societal factors are aligned in such a way that information technology can be deployed meaningfully. This is a difficult but essential task, since even great technology that goes unused is not particularly beneficial to anyone.

Data, information, and decisions need to be communicated accurately across the breadth and depth of manufacturing organizations. Many mechanisms can contribute to enhancing communication, including sabbatical programs for industrialists and academics in each other's territory, teaching factories, and advanced technology demonstrations that illustrate how the use of information technology can benefit factory performance.

Considerable research in social science will be necessary to facilitate the large-scale introduction of information technology into manufacturing. In particular, fully exploiting new technologies generally requires new social structures. Innovators will have to confront issues such as the division of labor between human and computer actors, the extent and content of communications between those actors, and how best to organize teams of human and computer resources.

Matters related to education and training will be central to 21st-century manufacturing. Given an environment of increasingly rapid change, continual upgrading of skills and intellectual tools will be necessary at all levels of the corporate hierarchy. "Just-in-time learning," that is, learning things as it becomes necessary to know them, may assume added importance.

Finally, although businesses depend increasingly on their intellectual and information assets, generally accepted accounting principles that businesses use to audit their finances and operations are derived from a business philosophy in which capital expenditures (i.e., expenditures that relate to the long-term value of a company) are associated with buildings and pieces of equipment. Research is needed to develop valuation schemes that appropriately account for the contribution of knowledge and core competencies to manufacturing and enterprise values.

2

Manufacturing:
Context, Content, and History

THE ECONOMICS OF MANUFACTURING

The economics of manufacturing is driven by the desire to produce salable finished products at as low a cost as possible while still maintaining acceptable standards of quality, functionality, and timeliness. Figure 2.1 describes in broad strokes the relationship between unit cost and production volume for three paradigms of production: manual, mass, and flexible. Flexible production has been the focus of recent efforts to apply information technology (IT) to manufacturing.

- *Manual production.* In manual production, the cost of producing an item is to first order independent of the production volume, since the dominant cost of production is the worker's time in producing the item.
- *Mass production.* In mass production, a substantial amount of capital is invested in a production line. However, once the facility has been built, the incremental cost of producing an additional unit is that of materials and labor, which is small compared to the initial cost of the facility. When the production facility is fully utilized, unit cost is minimized. However, such facilities by assumption produce a single product, require long lead times to deploy, and tie up large amounts of capital.
- *Flexible production.* Flexible production is still a goal rather than a paradigm. If successful, flexible production lowers both the capital and the time required to deploy a factory for a new product. Indeed, for certain types of products (e.g., integrated circuit chips, books) a "new" production facility is

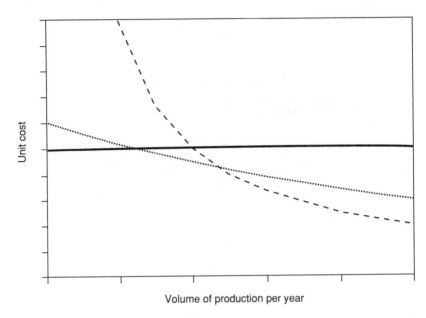

FIGURE 2.1 Unit cost versus production volume per year. Solid line, manual production; dashed line, mass production; dotted line, flexible production.

obtained from an "old" facility simply by making changes to the software that controls the production processes.

As Figure 2.1 suggests, manual production is superior to other types of production for those cases in which a highly customized product is needed in small volume *and* for which the nonmanual production of the product in a factory would require an expensive facility. When sufficiently large numbers of identical products are needed, mass production is generally superior. But the flexible production paradigm seems the most economical for intermediate quantities of moderately customized products that are needed in a timely manner.

THE NATURE OF MANUFACTURING

Manufacturing can be divided into two types—discrete and continuous. Continuous manufacturing refers to the production of substances or materials (e.g., the manufacture of chemical products). In continuous manufacturing, plant operations are reasonably represented by the well-understood mathematical formalism of differential equations. However, discrete manufacturing (e.g., the manufacturing of cars, airplanes, and other assembled products) is altogether different. Discrete manufacturing cannot be well represented by any known formalism.

For example, there is today no design formalism that automatically allows only a certain class of objects to be attached to another object (so that a handle but not a fork can be attached to a cup), although work on feature-based and constraint-based design tools is attempting to address this problem. The result is that predicting the operation of a discrete manufacturing plant is quite difficult and may be tractable only through the use of simulation models. The focus of this report is primarily discrete manufacturing.

Discrete manufacturing involves making discrete objects and is often based on actions such as shaping and assembly. The final product intended for the end user may be a desktop computer, a car, or a chair. But except in the simplest instances, the factory does not convert raw materials (e.g., sand, iron ore, wool) into a final product in one step. Instead, the final product is most often fabricated from components.[1] Each component is often itself the result of an assembly of subcomponents, and the number of steps between elemental raw materials and final product may be large indeed. Each component or subcomponent may be produced in-house or obtained from another supplier. Thus, in a sense, the "final product" of component suppliers may well be components for an assembler of end-user products. Another type of final product is created through the deposition of materials, in which a product is created by the selective layer-by-layer deposition of material on some substrate: both books and integrated circuit chips are created in this manner. Deposition-based production is also often used to create prototypes or product shells.

It is helpful to abstract manufacturing into four basic elements of an idealized process: product design, process design, shop floor production, and business practices:

- *Product design* normally begins with a combined effort by people who create new technology and people who meet customers to find out what customers need or want and what technology is available or feasible to meet those needs; this aspect of product design is often called conceptual design. In some cases, demand "pulls" and (less often) technology "pushes." What emerges is a high-level product model that usually contains a nontechnical statement of performance (e.g., provide instant communication with a distant computer) together with some quantitative goals (e.g., weigh less than 100 grams, communicate with a computer 300 meters distant).

Once a product's functionality is determined at this high level of abstraction, detailed design is undertaken to convert requirements into successively more detailed designs that anticipate or include, at each stage of the design process, the

---

[1] Fabrication is used here in a somewhat loose sense to denote both the mechanical connection of elements (assembly) or the conditioning of a component by the removal or treatment of some material, such as drilling a hole (shaping) or heat-treating (conditioning) a component.

implications for manufacturing methods, sales techniques, customer interactions, reliability and cost targets, field repair, safety, and environmental impact. As the product concept is refined, a variety of specifics may be added to the model, such as geometric details, materials or electrical specifications, and tolerances. Stylists, designers, and engineers make sketches, layouts, system diagrams, flow charts, and physical models as they try to define physically a specific product that will meet the stated requirements. Components and materials may be considered and rejected (because, for example, they weigh too much) several times before a promising solution is found. Cost, performance, and reliability must be predicted with increasing accuracy, as must potential manufacturing or use problems.

Conceptual design and detailed design interact. While it is clearly absurd to undertake detailed design before a designer has any idea of what the product is to do, feedback from detailed design may well influence the next iteration of the conceptual design.

A key element of design is verification. Verification activities test portions of a product design with prototypes or computational simulations, and test key portions of the overall fabrication process to ensure that cost, functionality, safety, and reliability requirements are met. Computer simulations allow the exploration of large numbers of test cases, but only actual physical testing can account for factors that cannot be adequately simulated; such factors range from small details (e.g., fatigue cracks) to major omissions (e.g., fundamental design flaws). Verification generates test results that can be used to improve product and process performance or quality.

The output from the product design activity is a model that describes the product with sufficient specificity and lack of ambiguity that it can be produced with a high degree of conformance to its specifications.

• *Process design* refers to the determination of an appropriate sequence of individual fabrication and assembly steps for converting raw materials and/or parts into a finished product. Process design is driven by a product's specifications and the processes available to produce the product. Process designers must find or create equipment and process plans (perhaps to be executed by others outside the factory) that will make and assemble parts into working products.[2] Process designers must ensure that each process step can be performed economically, accurately, and at the necessary speed. They must also ensure that the collection of steps, when executed, will result in a smoothly running factory. The output of this stage of manufacturing is a set of process designs that are compatible with the product design.

---

[2] The extent to which a process designer must develop a new production process for a new product depends strongly on the nature of the product being produced. As a rule, mechanical items require a greater amount of customized process design than do electronic items such as integrated circuit chips, for which manufacturing is mostly pattern-insensitive (see also Chapter 3, footnote 3).

- *Integrated product and process design.* Increasingly, product and process design are linked in an integrated effort. Although the steps above are described as though they take place sequentially, an integrated product and process design effort calls for them to be undertaken concurrently. Product design and process design interact through small feedback loops (whereby a change to a detailed product design may be made to simplify the process of making the product) and through large feedback loops (whereby feedback from customers may identify product design errors or ways in which process or product design can be improved). Such feedback loops are essential for uncovering incorrect designs or supporting data, incorrect assumptions about customers and their needs, incomplete specifications to suppliers, and inadequate tracking of performance in people or machines.

- *Production* implements the processes specified by the process designer. A qualified production facility ensures that these processes are capable of producing a product in sufficient quantity and of accceptable quality in a timely manner and within budget as often as needed. Moreover, qualification demands both technical assurances (e.g., that six-sigma defect rates[3] will be met for final products) and nontechnical assurances (e.g., that a supplier organization has sufficient financial staying power to guarantee a supply for a certain long period of time). Often working with the process designer, the production engineer makes decisions such as whether to make or buy a given part or process, what factors qualify a supplier, and how to manage engineering changes. In general, a production facility must provide for the receipt and acknowledgment of orders, the acquisition of materials, the performance of shop floor operations,[4] and the generation of the information needed to support continuous improvement.

A significant amount of real-time planning and scheduling is necessary to supervise those activities taking place on the shop floor. In addition, the production environment itself is complex and dynamic: machines break, resources such as parts or people are not always available, processing capability varies, communication falters, people with unique skills get sick or leave the company, and customer demands change. Consequently, the production facility must adapt by detecting changes, modifying intermediate goals, making trade-offs among conflicting goals, resolving constraints, and executing actions in a timely fashion. The production function also includes planning for machine requirements and resource capacity, systems management, and control. Production engineers must consider plant design, as well as identify inconsistencies and anomalies in the work contributed by product and process designers.

---

[3] Six-sigma defect rates correspond to a defect rate below 3 in 1 million.

[4] Equipment on the shop floor for discrete products performs two types of action: individual machines or work cells fabricate or assemble partially finished parts or subassemblies, and transfer systems move the output to the next machine or cell.

- *Business practices* are the aspects of manufacturing that go beyond the production activities occurring on the shop floor (the "touch labor" activities) and their real-time supervision. Business practices include marketing, designing and providing production facilities and equipment, managing materials and resources, procurement, contracts administration, financial accounting, and order taking. External suppliers must be found and their material and supply streams integrated into manufacturing operations for equipment or product components or materials that are not built or produced in-house. Planners must muster the capital to provide the necessary capacity to meet the anticipated demand for a product. Thus, manufacturing businesses care about the cost, quality, and schedule of all relevant processes, not just those related to what is traditionally associated with production (e.g., shop floor scheduling, machine utilization and control, and so on).

The steps within each element are not independent. For example, within product design, the specification of an object's geometry may be related to its functionality. Within process design, the particulars of how iron ore is converted to steel may affect how the steel is machined or shaped later. Within production, the placement of a particular piece of fabrication equipment may affect the speed with which a product flows through a facility. Rates of equipment utilization can be kept higher (tending to reduce costs) by a willingness to accept higher levels of inventory (tending to increase costs). Within business practices, suppliers must be chosen carefully to ensure that they deliver on time supplies that meet the required performance specifications. Orders must be processed accurately and converted quickly into work orders, schedules, materials purchases, and personnel assignments.

So also are the various elements of manufacturing interdependent (see cover illustration). Product design and business practices intersect in the market for products, the sources of finance, and the suppliers of components and equipment. Product design and process design appropriately linked constitute design for manufacturability,[5] while process design realized successfully in production is manifested as process excellence. Production capabilities significantly influence what product designers can create (although Box 2.1 illustrates another influence on product design). Finally, enabling and driving production tasks and business practices is the appropriate use of physical, human, and financial resources. The manufacturing philosophy of concurrent engineering is based on the interaction of all these elements; the manufacturing "wheel" depicted on the cover of this report is intended to suggest the integration of manufacturing activities both

---

[5] Indeed, today's product designers must concern themselves with a wide range of concerns that did not affect those of yesterday; these include product designs that facilitate manufacturability, assembly, repair, maintenance, recyclability, affordability, and so on.

BOX 2.1 The Relationship Between Product Design and Implementing Technology

A design engineer's choices about product design are necessarily constrained by the technologies available to implement those designs. Of course, implementing technologies include shaping and cutting tools and their capabilities for removing precisely controlled amounts of material. But a product also embodies technologies (e.g., the materials used in its construction, the technologies used to control its operation). When these change, the engineer's design options change—it is clearly a different design matter to build a glider out of wood rather than metal or plastic composites.

The technologies that control the operation and behavior of a final product are also crucial. In recent years, design options have been expanded by increasing capabilities to substitute electronic components for mechanical ones. Watchmaking is a good example. Fifty years ago, time-keeping accuracy was a direct function of the mechanical sophistication of a watch and the skill with which its components were fabricated, and high accuracy came only at great cost. Today, electronic chips and liquid crystal displays enable comparably accurate timekeeping at a tenth or a hundredth the cost of watches 50 years ago. Similarly, traditional mechanical fuel distribution in automobiles has been replaced by microprocessor-driven electronic fuel injection and ignitions, which have led to lower emissions, better fuel economy, lower maintenance, and lower-cost manufacturing.

An extreme case is illustrated in production processes that are based on the deposition and removal of material; both integrated circuit chips and books are produced through such processes. In contrast to a more traditional production paradigm in which raw materials are fashioned into parts and then assembled into products, fabrication by material deposition requires many fewer intermediate steps that are qualitatively distinct—repetition of the same process (with different parameters) is sufficient to manufacture the product. Those designing products to be produced in such a manner can generally operate with much more freedom in their work than those designing within a more traditional paradigm in which the fabrication of intermediate components constrains the design options.

Another example is the use of sensors incorporated into finished products. It is expected that sensors will be embedded into aircraft wing skins (for dynamic analysis and control of the aircraft), engine shafts (for torque readings), engine die castings (for thermal and strain information), structure members (for load and corrosion information), and tubing (for information on pressure, temperature, or electromagnetic radiation). Sensor-loaded products may be able to indicate where they are, how they are being used, when they have been damaged, or when they fail to meet a specified parameter. They may be able to monitor their environmental impact and signal for containment or destruction. Sensors may even enable a product to undergo "dynamic remanufacturing" even as it is being used. Alloys with "memory" are a crude example of a sensor-loaded product; Nitanol is an alloy that "remembers" its original shape when it is subjected to temperatures in excess of a certain critical value, and a deformed Nitanol structure will tend to revert to its original configuration upon heating. A more active example is provided by recent work at the Xerox Palo Alto Research Center on active control of structural elements. Work there has demonstrated that by using a sensor and a computer to sense and compensate for initial deflections in a support column subject to buckling stresses, the strength of the column can be increased by a factor of 4 with a negligible increase in weight.

among the four basic elements of manufacturing (the four spokes of the wheel) and across these elements (between the spokes). In contrast to traditional product development, which involved little communication between the various elements of manufacturing, concurrent engineering emphasizes improved communication and early consideration of "downstream" issues such as the relationship of product repairability by the end user to the ease of assembling the product. Concurrent engineering stresses more or less simultaneous consideration of product and process design, customer needs, and business practices in order to speed the development process and avoid costly errors or redesigns.

THE HISTORICAL CONTEXT OF MANUFACTURING

Change is ever present in manufacturing. New technologies brought to the marketplace in the form of products and capabilities become part of the platform that manufacturers use to create the next generation of products. New economic conditions, new knowledge about how to organize manufacturing effort, new consumer preferences, and new manufacturing techniques constantly alter the manufacturing front. In a few decades, manufacturing has undergone a major change with respect to the environment in which it operates, the methods through which it conducts business, and the technologies that support it.

Early Paradigm Changes

Manufacturing has undergone many paradigm shifts from the Bronze and Iron Ages to today's technology, but the change has been especially apparent in the last few hundred years, affecting the way manufacturing was performed, the processes used, the products made, and the economic power of the locale in which manufacturing was taking place.[6] These paradigm shifts have ranged from Eli Whitney's invention of interchangeable parts to the invention of numerically controlled machine tools. In each case the paradigm shift has resulted in an increase in productivity of about a factor of 3 beyond the old method (Figure 2.2).

In the earliest paradigm, the transformation from raw material or subassemblies into a more valuable final product was carried out by skilled artisans and craftspersons, people who practiced under expert supervision until they achieved proficiency. These experts performed the entire task of transformation, from raw material to final assembly and test, mostly by hand. An area's economic wealth depended on the skills of local artisans and craftspersons, and world fame accrued to specific areas that manufactured specific goods. London, Imari, Leeds, Birmingham, and other artisan centers achieved world renown.

---

[6] A more comprehensive, engineering-oriented examination of manufacturing epochs can be found in a table on the evolution of manufacturing in Jaikumar (1988).

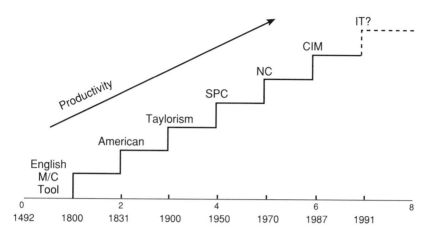

FIGURE 2.2 Productivity follows paradigm shifts. The solid line in this figure from Jaikumar (1988) illustrates how each paradigm shift has resulted in greater manufacturing productivity. The dotted line was added by the committee to suggest that the increasing use of information technology (IT) to integrate manufacturing activities may have an impact comparable to those of previous changes in manufacturing. SPC, statistical process control; NC, numerical control; CIM, computer-integrated manufacturing.

During the Industrial Revolution, when steam power became readily available, economic wealth shifted to locations that had inexpensive access to power sources such as coal, oil, and hydroelectric power; to raw materials such as iron, aluminum, and copper; or to low-cost transportation via rivers or seaports. Economic wealth was determined largely by the capital equipment available to transform raw materials into finished goods. Pittsburgh, Gary, the Ruhr Valley, and other smokestack areas became centers of the new manufacturing capabilities.

The latter part of the Industrial Revolution introduced mass production methodology, changing the nature of work from a "do it all" process to a specialization process. Specialists now performed repetitive tasks in one specific activity, substantially decreasing the cost of the finished product. Interchangeability of parts became critical, but knowledge about how the entire product came together decreased. Detroit, Wolfsberg, Osaka, and other cities became centers of mass production.

One effect of these paradigm shifts, as shown in Figure 2.3, is that employment in the United States in the manufacturing sector has dropped drastically even as the value of goods shipped has remained relatively constant as a percentage of gross national product.

The next major paradigm shift occurred after capital, manufacturing technology, and access to raw materials became widely available and no longer represented competitive advantages; U.S. manufacturers faced greater, worldwide

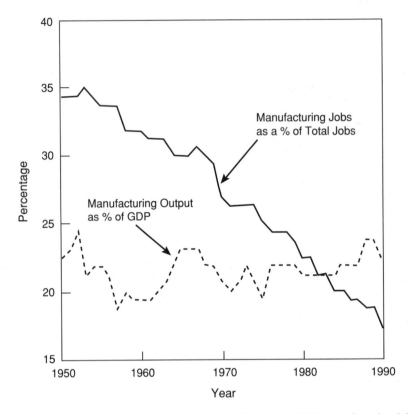

FIGURE 2.3 Manufacturing over time—relationship between U.S. manufacturing jobs and manufacturing output, 1950 to 1990. GDP, gross domestic product. Courtesy of Bureau of Labor Statistics.

competition. Various approaches to reducing costs and improving delivery were undertaken, with significant attention being paid to industrial engineering. Knowledge became important, as did quality control, time studies of manufacturing processes, flexible organization, skilled workers, and so on. The availability of an educated work force became a driver of economic wealth. Silicon Valley, Los Angeles, Seattle, Tokyo, Route 128, and others became the new centers of excellence.

Recent Changes and Their Effects

From the U.S. perspective, perhaps the single most profound change in the manufacturing environment in the recent past has been the emergence of a competitive worldwide market in sophisticated manufactured products as other nations have capitalized on advantages not available to the United States (e.g., inexpensive yet highly educated labor, cheap capital). In the early stages of this evolution, quality disciplines and process control methodologies became competitive weapons. Japan, with its national quality programs and intense focus on quality, was the first country to use this strategy widely, but others have followed suit. The global pursuit of quality has evolved to the point that quality, in itself, no longer confers a significant advantage: it is simply taken as a "given" by consumers everywhere.

Product quality, more rapid delivery, better asset control and utilization, and the ability to execute more complex manufacturing tasks and build increasingly complex products have come to characterize excellent manufacturers. Manufacturers are now linked directly to their suppliers and customers. Many retailers collect worldwide sales data every day and modify their suppliers' schedules in response. Products are designed to suit regional styles and needs, even if they are made in other regions. International payments are made in a variety of currencies as materials are purchased, labor is obtained, ships are laden, and products are transhipped. These business and marketing issues are not usually associated with the more technical concerns of manufacturing, but in fact they are central to its success and enlarge the very definition of manufacturing. Most importantly, they are very information-intensive. Without information-driven links to financial markets, logistics services, and market knowledge, manufacturing businesses would operate in a vacuum or seek blindly to force their output on unwilling customers and ultimately fail.

3

Integrated Product and Process Design

INTRODUCTION

Information age manufacturing begins with information age design. Designing both a product and the processes by which it is produced involves understanding what the product is to do and how the product will do those things, converting the requirements for the product's behavior into engineering specifications, and producing plans that marshal the materials, equipment, and people needed to make and deliver the products. Even apart from the pressures for shorter time to market for new products that stress current design paradigms, new design challenges are generated by new product trends (e.g., shrinking feature size, decreasing tolerances, or a growing numbers of parts) and by new manufacturing processes (that designers must learn to exploit).

As a result, the amount of knowledge and data relevant to product and process design is rapidly becoming more than a single individual can comprehend. A further complicating factor is that an integrated product and process design (IPPD) effort must usually be coordinated among a number of engineering teams with different specialities and from different companies, since it is rare for a single company to have all the skills, technologies, and financial resources to design in-house all of the components needed. Managing this coordination task represents a major opportunity for information technology (IT) to have a positive impact.

As described in Chapter 1, a comprehensive IPPD system would include integration of performance specifications, conceptual design, detailed design, manufacture, and assembly, together with the ability to simulate actual use, field repair, upgrade, and disposal. Such a comprehensive system will not be possible for many years.

However, an important first step toward this vision can be achieved by joining detailed design with manufacturing and assembly. To accomplish this requires a new level of information structuring and integration. Feature-based design is the best way currently known to capture and integrate the necessary information that links the geometry of parts with their functions, fabrication, and assembly. Present computer-aided design (CAD)[1] systems support creation of geometry only. Apart from stress and thermal analyses of parts and certain types of kinematic analyses, most design analyses must be done manually because there is no way to obtain the necessary information from the circles and lines stored in the CAD system.

An IPPD system that realizes this first step toward a more comprehensive system will consist of three elements: a database, a set of algorithms, and user interfaces. The database will be structured to capture the information about the design in the form of geometry plus features (places of interest on each part, together with information on what role they play in the product's function and how to make and assemble the features in relation to each other). The algorithms will take the information they need from this database to simulate function, determine optimal assembly sequences, estimate fabrication or assembly cost, or perform design-for-assembly analyses, for example. The user interfaces will make it easier for the designer to create a design using features and apply the algorithms to study and perfect the design.

Prototype software that does some of these things exists now. This software can be a basis on which to build a new kind of computer-assisted design that integrates technical, business, and economic issues relevant to design. It can support analyses of cost and function, as well as the study of families of products that share parts or subassemblies. Such software has been demonstrated for the design of certain complex electro-mechanical items.

Another IT-enabled connection between design and manufacturing is the use of stereolithography as a visualization aid for designers and as the basis for rapid generation of prototype molds and dies for the production of mechanical parts. In an experiment conducted by a major automobile manufacturer, vendor bids based on a drawing and a stereolithographed model were lower than bids based on a drawing alone; this result was explained by the fabricator's greater ability to visualize the complexities of the item in question and thus to more accurately determine the costs of its fabrication.

---

[1] Over the years, the acronym CAD has evolved in meaning. CAD initially stood for "computer-aided drafting"; however, perhaps because information technology achieved greater penetration into the world of design engineers, it has come to mean "computer-aided design," of which one part is drafting.

Once a core design system is in trial use, it can be extended up and down the design process to include concept design and field use considerations. The information hooks in the data structures will be there to integrate with functional simulations, repair environments, and other aspects of product design. This evolution of the design system will be fueled by experience with its core implementation; feedback from users will determine what new capabilities it needs.

DESIGN PARADIGMS

Electronic Design

Of all the different types of design for discrete manufacturing, the design of very large scale integrated (VLSI) circuits (described in greater detail in Box 3.1) is today the most sophisticated. Thus, it is reasonable to suggest that the VLSI

BOX 3.1 An Example of Information Technology-driven Integrated Product and Process Design

In the semiconductor industry, electronic computer-aided design (CAD), also called electronic design automation, is a good illustration of how information technology can support the integrated product and process design process. Today, a designer of a chip using commercially available CAD tools would undertake the following steps:

• *Conceptual design.* Customer requirements are determined and converted into design goals (e.g., instruction set, power consumption, and chip size and speed). Designers choose an underlying technology based on these goals. In addition, an associated "framework" orchestrates the operation of various other design tools, thus enforcing some of the methodology (e.g., cannot proceed to step *x* before step *y* is complete).

• *Detailed design.* In detailed design, the conceptual design is reduced to logic elements that take in, convert, and put out logical values. A designer determines the basic architecture of the chip (how it will be partitioned into functional units) and then writes a description of each functional unit using a hardware design language such as VHSIC Hardware Design Language (VHDL). Detailed design involves making trade-offs among power, speed, and size; tools for detailed design address automated logic design, test vector generation, formal verification to ensure that the gate-level view of the chip matches the VHDL description, early timing analysis, and the spatial arrangement of various components. Tools are also available to perform a variety of detailed design tasks, such as static timing analysis, partitioning, layout, routing, wiring, and circuit-level timing; results from these tasks may impel the designer to alter the VHDL description. Unless a chip is to be fully customized (perhaps for reasons of maximizing speed of operation or minimizing power consumption or size), libraries are used to provide logic components to be integrated into the new chip. Detailed design results in a set of chip masks that drives the production process.

• *Simulation and verification.* The detailed chip design must be tested to determine correctness and the extent to which the design meets functional requirements. *Functional* simulation ensures that the chip, as represented through the VHDL description, actually meets customer requirements for logic. *Physical* simulation, orders of magnitude slower than functional simulation, takes place at the gate or transistor level to ensure that timing and other layout-dependent issues are handled appropriately. Physical tests are performed on the chip that may reveal flaws in the simulator used to design the chip, flaws in the fabrication process, or design flaws that were initially unnoticed by the designer.

The design stages and simulation tests are repeated until results suggest that the overall chip design is adequate. Process design generally need not be undertaken to create a fabrication process for a particular chip, since chip designers design only products that a given factory can fabricate; process design is needed to plan the factory and the fabrication technologies the factory will use to produce electronic chips of near-arbitrary design within one product or process technology family.

The design of a semiconductor technology requires similar steps. What results is a set of design rules for the technology and high-level chip masks (used for etching the chip's features onto the semiconductor substrate) and usually a set of tools and library functions that the designer can use in implementing the logical and physical design of a chip built with this technology. These library functions consist largely of Boolean functions, circuit components such as latches, registers, register files, and memories, and design rules that define how they can be connected together in a logic network. Each library cell has a geometric view, a functional view (e.g., a three-input NAND gate) for use in logic synthesis, a behavioral view for use in high-level simulation, a fault view for use in testing, and a timing view for use in timing and delay analyses. Cell libraries may contain families of same-function circuits with different timing, power consumption, test characteristics, and geometry.

design paradigm might suggest directions for improving the IPPD process in other manufacturing domains. An examination of VLSI design suggests that its successes are based on certain key attributes, including:

• *Appropriate design abstractions.* The amount of detail needed in each step of electronic design is appropriate for the work to be performed at that step, thus allowing, for example, the three-dimensional characteristics of a chip to be virtually ignored until the final step of the detailed design process. Central to these design abstractions are languages that describe electronic products at many levels of abstraction (e.g., Verilog HDL) and the ability to use these descriptions in simulation, in design-space exploration, and in automation of a large part of a product's detailed design. Further, design tools can operate at different levels of abstraction and still obtain feedback from lower levels in the abstraction hierarchy; such tools enable an iterative refinement of the chip design that starts with high-level estimates and continues to the manufactured chips. Today, high-level chip specifications allow design alternatives to be explored almost at once, leading to early detection of conceptual-level errors and pruning of unprofitable paths. The early design can be refined and made more specific—data collected from this level can be back-annotated and used by higher-level tools.[2]

• *Appropriately parameterized "parts" that can be used in the design process.* Chips with new designs often make considerable use of logic functions that have been implemented before, for example, Boolean functions, latches, drivers, and receivers and terminators. Libraries of these basic functions (called library cells) and assemblies of these cells (called macros) that implement more sophisticated functions enable a designer to use prevalidated designs (perhaps with the specification of a few parameters) when appropriate, with the result that these functions are more likely to be integrated into a new chip correctly and in less time than if they are designed from scratch. (On the other hand, since connecting predesigned building blocks almost always does not result in optimal chip perfomance, chip designers may choose to forego the advantages of libraries to improve various dimensions of chip performance.)

• *The ability of the designer to ignore most off-nominal behavior and side effects in the design process.* In practice, the behavior of transistors is influenced very strongly by their context. For example, a logic gate generally has different speed and power characteristics depending on where it is used. However, the design rules associated with library cells and macros and the tools that enforce adherence to these rules enable a designer to treat library components as parts that can be connected without concern for issues such as back-loading.

• *A close relationship between the fabrication processes and the extent to which the actual product matches the original design.* If a designer adheres to the set of chip design rules[3] appropriate for a given semiconductor technology, a properly controlled fabrication process will yield the desired product and desired product behavior (even if individual chips are different at some level). "Control" usually includes scrupulous attention to impurities in materials, surrounding

---

[2] A specific example of this iterative design process is the use of logic synthesis and early (prephysical design) timing. Statistical methods are used to time the logic before it is placed and wired, and automated logic design systems (logic synthesis) can use these data to tune the logic design to meet timing requirements. After physical design, the timing is refined and made more accurate because more precise information about wire lengths and routes is present. The logic can be annotated with the exact times and be brought back to logic synthesis for readjustment of the logic design based on the new information.

[3] The fabrication of digital chips uses what are essentially fixed manufacturing processes that are configured by patterns (plates and masks) derived directly by algorithms from the chip specifications. For this reason, chip fabrication is sometimes called a pattern-insensitive manufacturing process. The role of design rules is to restrict the patterns to those that will be faithfully honored by the fixed production process.

gases, and process fluids and is pursued to ensure that the process itself remains stable and thus able to honor any design obeying the design rules.

While the chip design process in actual practice generally requires more manual intervention than the idealized description above would suggest, most experience in electronic design suggests that automated design tools have dramatically increased the complexity of chips that can be fabricated and have reduced the time needed to deliver designs of constant complexity.[4]

Application to Mechanical Design

Mechanical design poses myriad different problems, and the extent to which the electronic design paradigm can be applied to mechanical design is a matter of some debate. Digital logic can be considered as a special class of product whose design and fabrication problems have proven amenable (with the expenditure of considerable R&D resources) to the application of information technology. By contrast, mechanical items represent a wholly different class of products for which there is currently no formal representation of function and there is no direct algorithm-dominated way for reducing a functional description to a physical design; whether this is fundamentally true or merely a limitation on current knowledge is as yet unknown.

Most importantly, the science and engineering underlying models of mechanical products and the processes to manufacture them are not nearly as well understood as those for electronic products. Mechanical designs are characterized by complex and large multimedia energy interactions between a limited number of elements and by changes in element behavior over time. The information needed to describe multiple behaviors of mechanical systems is difficult to express in a single format or language, and there are few tools for representing or

---

[4] This is not to say that electronic design does not have limitations. Moore's Law states that the number of transistors on a state-of-the-art chip will double about every 12 months. However, the design capability of engineers is not increasing with time nearly as fast as the number of transistors. Using today's design tools, an expert electronic designer engaged in fully customized design can reasonably hope to complete the necessary design work on a few hundred transistors in a single day of work; such work includes the detailed design, debugging, and documentation needed for a commercial product. For example, the Pentium processor (order of magnitude 5 million transistors) took about 100 engineers 2 years to develop, a rate that corresponds to approximately 250 transistors per day per engineer. The resulting gap between the complexity of a chip and the design capability of individual engineers must be filled by the use of additional engineers. Over the long run, two approaches for filling the design gap seem plausible: better tools to increase the design capability of individual engineers, and the reuse of existing and validated designs to reduce the amount of work that must be undertaken from scratch. Some designers today claim the ability to lay out hundreds of thousands of transistors (tens of thousands of gates) per designer-year by using high-level design paradigms with tools and programmable chips.

BOX 3.2 Aspects of the Mechanical Design Process

Three basic elements of mechanical design are the following:

1. *Conceptual design.* The design of mechanical systems often calls for the management of significant energy flows, the consideration of complex three-dimensional geometry, and the understanding of relationships between geometry and energy. The "underlying technology" in which a mechanical system will be implemented is often not naturally specified by the product requirements, and the space of design goals is much more multidimensional than in the case of very large scale integrated (VLSI) systems.

2. *Detailed design.* For mechanical items, detailed design is generally independent of conceptual design, since no formal methods exist for linking concepts or functions to detailed geometry except in perhaps a few special cases. In most instances, detailed design of mechanical systems that are even marginally efficient in their use of space, weight, or energy generally requires the custom creation of integrated combinations of three-dimensional geometry rather than piecing together predesigned and tested library building blocks. Using standard parts or designs for parts used for other products is not easy because geometries differ greatly even for the "same" item, and there is no way to catalog them systematically. The few standard library or catalog items that are available are generally not main function carriers but instead are fasteners, bearings, motors, valves, and other similar items. Difficulty in mechanical design often centers on creative geometric reasoning, management of multiple behaviors, mitigation of unavoidable side effects, and anticipation of a variety of failure modes. (In some cases, a mechanical design is an evolutionary outgrowth of a past design. In these cases (such as automotive suspensions) progress is being made in establishing design templates that capture in parametric or rule form the traditional parts and relationships among parts that every good example of the genre must contain. Design then consists of packaging the given elements, using existing simulations to confirm basic behaviors.)

3. *Verification.* Computer simulation tools can predict nominal behavior at the

---

testing mechanical designs that are comparable in power to those available for the analogous electronic design task. A similar point applies to the modeling of multiple and simultaneous high-level energy interactions. Compared to the problems encountered by the VLSI circuit designer, the problems that challenge the mechanical designer, described in Box 3.2, suggest important research questions for better IT support of the design effort. These challenges are the subject of the remainder of this chapter.

NEEDS AND RESEARCH FOR MECHANICAL DESIGN

As noted above, it is not clear that the paradigm of electronic design can be applied to mechanical design, the area that the committee believes poses the greatest need today. Nevertheless, the successes of today's electronic design paradigm suggest research areas to improve mechanical design.

system level if only one or two main modes of behavior are simulated at once and only in the case of sufficiently simple mechanical systems. Since side effects and off-nominal behaviors often cannot be adequately modeled and usually cannot be designed out, either extensive, time-consuming, and costly prototyping and field testing are required, or the designer must design the system very conservatively to mitigate the consequences of these side effects to an adequate degree over the expected life of the system.

Process design is integral to the design of individual mechanical systems. One reason is that for many mechanical parts, there is no way to automatically convert product geometry and other specifications to a fabrication process such as a sequence of machining steps. Designing fixtures, for example, is a major problem. Moreover, most mechanical fabrication processes are neither pattern-insensitive nor precisely controllable; even when the commands to a given process are repeated identically, the output of the process (as implemented on an actual production line) is different each time, and the differences may well matter. A command to drill a hole 0.500 inches in diameter may be provided, but the resulting hole may be 0.499 inches in one case and 0.502 inches in a second case. In general, elimination of these differences is either too difficult or too expensive. Thus, process design involves the choice of an economical process whose output comprises mostly acceptable (though different) parts. In most cases, there are no simulations that can predict the range of variations in nominally identical outcomes that a given mechanical process might generate. This range, usually called "process capability" when appropriately normalized, is often estimated by experienced people. (Such "tolerance-like" problems also affect chip fabrication: for example, a chip that emerges from the fabrication process with line widths that are too large or too small may run more slowly or not at all. Such issues affect yield rates dramatically. The difference is that for electronic fabrication, the process that determines yield is independent of the pattern on the chip to a considerable degree.)

The above three steps may have to be repeated until a functional and manufacturable design in achieved.

Specifically, an important goal is the development of appropriate abstractions at every stage of the design process and tools that manipulate these abstractions. Abstraction is not the same at each phase of a design, and resolution of implementation details is deferred in the design process until such details are really needed. The use of such abstractions reduces work because existing representations can be reused, and different representations can be related through their common pieces. Key to developing these abstractions is the exploitation of hierarchical composability (i.e., the construction of complex standards or information models from simpler ones). Researchers should seek appropriate abstractions for mechanical products and develop tools to support design in terms of those abstractions.

In electronic design, many sophisticated tools have been developed to support physical aspects of design, while fewer tools exist for aspects of conceptual design such as requirement gathering or architectural decision making. This disparity suggests that it is far easier to collect, characterize, and represent data about "things" than about ideas and decisions. Nevertheless, the payoff for automated support of activities such as requirement gathering, corporate decision making, and overall product architecture is so high that research directed at these targets is also worth undertaking.

The committee believes that concentrating on the areas described in the rest of this chapter will yield the most significant benefits to the IPPD process in the short to medium term. Other areas in the IPPD process not described in this chapter (mostly in the conceptual design area) will be advanced in the short to medium term by better communication capabilities and by better and additional access to data.

The committee notes also that success in improving design is likely to yield particular benefits to small manufacturers. To a considerable degree, tools that support product and process design can be categorized as "mostly software"; that is, they require low capital investment to obtain. Moreover, the increasing power of computational hardware (and its dropping cost) will make these tools more accessible to manufacturing firms with small capital budgets. To capitalize on this enabling trend, it is necessary to design these tools and their interfaces so that small businesses with limited technical depth can use them readily. A growing information infrastructure (Chapter 6) will link these small firms with each other and with larger firms. The emergence of standard data formats and interoperable tools will facilitate the spread of advanced capabilities.

Table 3.1 lists several areas for research in product and process design. In general, the research proposed by the committee focuses on design-space exploration, creating (parameterized) geometry, characterizing tolerances, predicting failure modes, increasing robustness, and facilitating reuse of components and designs. In addition, major issues of scale-up and complexity remain to be addressed explicitly as an integral part of any research done in these areas. Explaining these research foci is the subject of the remainder of this chapter.

TABLE 3.1 Research to Advance Product and Process Design

Subject Area	Example of Research Needed
Multiview design descriptions	Design by function Relation of geometry to function Functional simulation Parametric design
Capture of nominal and variant behavior of products and processes in one model	A mathematics of variation for performance modeling Descriptions of product function and variants directly related to descriptions of geometric or material variations

TABLE 3.1 Continued

Subject Area	Example of Research Needed
Multipurpose data and model representations	Techniques for relating models at different levels of abstraction Logic-based representations Protocols, formats, and representations for data interchange among models
Design methods and tools for groups of parts and systems	Decomposition methods to break product concepts into subsystems Subassembly performance models and interface descriptions for joining subassemblies to each other Assembly planning Trade-off analysis (e.g., cost and design)
Process description languages and models	Set of process primitives (building blocks) from which process models can be built Languages with syntax checking for correctness and completeness of process descriptions Resource description models Accommodation of spatial and temporal dimensions of processes
Novel design considerations	Easy, error-free configuration control at the selling or servicing stage Manufacturing of a robust final product from parts obtained from different sources
Product-process data model	Data descriptions for many physical processes and entities in a unified form Descriptions of design interactions, analyses, and process steps integrated with product geometry and function descriptions
Decision aids	Data visualization Database searching using geometric features, performance criteria, or process descriptions Intelligent advisors
Geometric reasoning	Visualization tools
Knowledge and information management	Systems that capture corporate memory and knowledge Systems that support corporate learning Techniques for handling data legacy issues Systems that record design history and rationale

Research for Product Description

Communication among engineering professionals has always relied on models—sketches, drawings, analytic models of behavior, or many other symbolic representations of knowledge related to products. However, what distinguishes data or information models from arbitrary documentation, such as simple reports or drawings, is the addition of formalization.[5] This formalization is motivated by the need for unambiguous communication between collaborators and/or our desire to use and interoperate among a growing set of computer-based application tools. Specifically, a data model is expressed in some data description language. A language specification defines the form and meaning (syntax and semantics) of entities in the language. The initial graphics exchange specification (IGES; *IGES*, 1993), for instance, specifies the language of IGES data files by specifying their syntactic form and relating the entities in the data files to known geometric entities.

Traditionally, the specifications of data description languages such as IGES have been written in a natural language such as English, and a human being must use that description to build a software parser/recognizer for the data description language. The semantics of the elements of the language remain expressed solely in English. More recent efforts such as PDES/STEP (product data exchange using the standard for the exchange of product model data; ISO, 1994) provide a more formal language, EXPRESS, for use as a data description language in which data models are described. EXPRESS allows the modeler to capture some of the semantics of the data by explicitly recognizing relationships between data elements along with the cardinality of such relationships and by capturing constraints between data elements. Box 3.3 describes two approaches to knowledge representation. Appropriate formalisms also support the generation of agreement on queries about data; application tool development; language translation; and services such as change notification, management of information dependencies, and matching of information producers and consumers.

A highly general expressive capability is needed to support the exchange of

---

[5] The term "formalization" is used in this chapter in the sense of "a greater degree of rigor, clarity, and explicitness that would be necessary for computer-based representation, manipulation, and analysis." However, the extent to which it should include matters such as formal proofs of completeness or correctness is subject to some debate within the community, with some advocating much higher degrees of mathematical formalization than others. Those who believe in high degrees of mathematical formalization tend to be logic theorists. The advantage of logic-based approaches is that they are neat and clean and amenable to the power of formal logic and mathematics. On the other hand, it is not necessary to embrace this degree of formalization in an attempt to get away from models specified in English and a few drawings. Indeed, the need to cope with the uncertainty of the manufacturing environment and with spatial and temporal relationships (for example) requires that a formal logic be modified in a way that reduces the "neatness" of the formalism and its ease of use as a base language for representing interesting phenomena. The result may be a more "scruffy" approach to the problem.

BOX 3.3 Knowledge Representation Foundations

Different computer-aided tools for working in different domains may require different data models for a given product. Nevertheless, since such models generally contain much common structure that can be exploited to reduce the work in creating new models and to provide the basis for generic tools that operate on representations of knowledge, it is inappropriate to create separate, top-to-bottom standards for each domain of interest.

The first requirement is a basic framework that provides building blocks sufficient to our diverse forms and uses of product knowledge. Once these building blocks are in place, it is possible to develop standards for domain-specific applications. Two approaches to developing the base building blocks appear promising:

• *Information modeling languages,* such as the EXPRESS language that is used to state the PDES/STEP standards, which provide a capability to specify machine-readable definitions of shared data. EXPRESS allows a designer to describe object classes with attributes and inheritance, which offers limited formal semantics for the data (i.e., for database operations). The PDES/STEP effort has encouraged the development of tools that can operate on the EXPRESS language specifications to provide capabilities such as graphical display of data structures and generation of code to support tool creation or database implementations.

• *Knowledge-representation languages,* such as the Knowledge Interchange Format (KIF). Knowledge representation languages are based on the concept of a "shared ontology," which is simply an agreed-upon set of terms and meanings that enable two parties to exchange knowledge for some purposes in some domain. In practice, it is a dictionary of classes, relations, functions, and object constants, along with their definitions in human-readable text and machine-interpretable sentences. This approach offers enormous flexibility, including the taxonomic capabilities of object-oriented and frame-based approaches, and the ability to construct axiomatic theories. Consider, for example, the following query posed by an analyst working on a product design: "I added the following assumptions to theory Z . . . and took the following quantities as given . . . . I then arrived at the following conclusion (part X will fail). Do you support my conclusion?"

Because they are rooted in first-order logic, the knowledge representation ontologies are more expressive and hence more flexible than EXPRESS-based STEP standards. Conversely, the STEP standards provide much more direct support to developers of computer-aided applications through application protocols and other mechanisms provided in the standards. Translation facilities between EXPRESS- and KIF-based constructs allow complementary use of both formalisms.

---

engineering knowledge, specifically the ability to exchange a conceptualization that specifies the objects that are presumed to exist in some area of interest, as well as the relationships that hold among them (Genesereth and Nilsson, 1987). For example, for two parties to discuss models of dynamical behavior, they must first agree on the use of terms such as generalized coordinates, position, and velocity, and they must understand the laws on which the models rely (e.g., Newton's laws, Lagrange's equations, or Kane's equations). At the same time,

---

BOX 3.4 Domain-specific Libraries

In creating domain-specific libraries of representations, a balance must be struck between short-term needs (dictated by existing tools) and possible future uses of the knowledge. Tool designers must also accept the reality that both the tools and their context for use will be changing continuously as technology and organizations evolve.

Any specific modeling choice constitutes a compromise between a variety of competing interests—the chief trade-off being between specific or short-term needs and generality. For example, a controls design department and a dynamic analysis department in an organization may regularly exchange models of a satellite's dynamic behavior. But the information models appropriate to their needs may require a simple model that accounts for only a single degree of vibrational freedom or a complex model based on partial differential equations. At a minimum, it is necessary to identify the intersecting knowledge in the two views that allows a more formal bridging of the gaps between their different views of the world.

Standardization, or the practice of conforming to a particular representation, is an important part of sharing knowledge efficiently. Standardizing at the level of knowledge representation greatly reduces the overhead costs associated with translation and the construction of information models. However, the need for large numbers of domain-specific information models suggests a very flexible approach to standardization. Unlike choosing between syntactic variants such as IGES and DXF,* there is much more to distinguishing among different representations of a behavior theory like kinematics. In kinematics, a variety of formalisms exist that can model the same behavior (e.g., quaternions, matrix exponentials, screws, Euler parameters, Euler angles). Even particular representations of screw theory could differ significantly in their completeness and in the manner by which basic entities are defined. Individual information models, then, will be evaluated by potential users in much the same way that designers evaluate the choice of motors for a particular device. Models will come in many shapes and sizes with varying capabilities; some will be good, some bad. A wide variety of competing representations is expected and even desirable.

---

*Design eXchange Format, a proprietary interchange standard controlled by AUTOCAD, San Raphael, Calif.

---

general-purpose approaches, while inherently flexible, are intrinsically clumsy to use precisely because they do not provide specialization to individual domains of interest. Domain-specific approaches may be more efficient for use in those individual domains and may be more expressive for them (Box 3.4).

Examples of the research needed for product description to support design efforts include:[6]

---

[6] Additional discussion of these examples can be found in MSB (1988), pp. 17-19. It recommends research on product and process design, including data structures for describing products in terms of conceptual design, functional features, dimensions and tolerances, manufacturable features, and so forth, and methods that allow such structures to be interfaced with other computer-integrated manufacturing components, such as knowledge-based systems.

• *Design descriptions that allow multiple views of a product* (geometry, engineering parameters, functional requirements and behaviors, and relationships among these domains). Data models should be capable of representing the product or component from many different views at different stages of design, manufacture, and use. The product designer should be able to view its geometry, tolerances, assembly problems, and repair scenarios; the techniques of feature-based design and parametric design hold particular promise (Box 3.5). The current PDES/STEP effort does address some of these issues; for example, an addition to the STEP Application Protocol called Configuration Controlled Design will contain both the geometry of a product and information on the revision, release, and effectiveness associated with producing this product. Nevertheless, no adequate data model exists, nor are there adequate methods for supporting these activities.

• *Model formulations or simulations that are capable of describing the main off-nominal behaviors, or variants, as well as the nominal behavior of a product.*[7] For many products, more effort goes into anticipating and mitigating a wide range of off-nominal behavior (e.g., testing aircraft for fatigue and crack resistance, determining if layers in a microprocessor will peel apart under temperature extremes) than into determining how to meet the main functional requirements. (See Box 3.6.)

• *Better understanding of relationships between different representations and models.* Because of the interdisciplinary nature of collaborative product design, relationships between different representations are important. Questions such as how to relate a model in someone else's view of the world to a model in one's own view are ubiquitous. Determining relationships between models (e.g., between models of different levels or types of abstraction) will remain a difficult problem for some time, but a farsighted approach to knowledge representation can greatly enhance computational support through the exchange of information among different applications. Logic-based representations may help in this area in the future, although there is debate on the point; further exploration is needed.

• *Formats and representations that enable data exchange among different design tools.* While different computer-aided tools for design are based on different data models, much of the content of these models is overlapping. As more computational support for design comes on-line, engineers will rely

---

[7] A 1991 NRC report recommends research on tolerance analysis, tolerance representations, tolerance-performance relationships, and tolerance standards and measurement methods. See MSB (1991), pp. 56-57.

BOX 3.5 Feature-based Design

A feature-based data model is a way of describing a design that contains more information than just geometry. In a computer model, a circle may represent a hole, but it is not a hole. However, feature-based data could include text data saying that it is a hole and giving the diameter and tolerances, plus numerical control instructions for how to drill it. The "hole" is then called a feature, which in fact is a data object that contains instructions for how to draw itself on the computer screen and how to drive a machine tool to drill it. Features can describe machined areas, or they can describe areas where a measurement will be taken ("measurement features") or where one part will be joined to another ("assembly features"). Features are thus able to hold a great deal more information than a purely geometric model can. In principle, features can contain design intent as well as details about how to make and use the feature. For example, a pocket to hold a precision ball bearing would require tighter tolerances and a finer surface finish than a hole through which oil is squirted. Feature-based models may also reduce storage requirements, because compact functional descriptions could be stored and voluminous geometric descriptions could be generated from the functional descriptions when needed.

Feature-based design also supports higher-level product data models and descriptions of configurations. A simple model of configurations may include relationships among subparts and attributes. Such a model can be used to create a more specialized theory of configuration design that includes connections between parts, special part subclasses, predefined lists of available parts, and so on. The idea is to create multiple layers of representation that mimic the multiple levels of abstraction at which we view things. This approach has two benefits: work is reduced through reuse of existing representations, and different representations can be related through their common pieces.

Several STEP projects provide for specification of simple form features, for example, through and blind holes, although the taxonomy is far from complete. In addition, new work has just been initiated in providing a parametric representation within STEP, an ability that is necessary to support more extensive use of form features. Ford Motor Company is actively working on a research project called Rapid Response Manufacturing that makes extensive use of form features. Output from this project is expected to drive development of the STEP standard in this area. This program is being conducted jointly with General Motors, Texas Instruments, United Technologies Corporation, and Allied Signal.

increasingly on information models to help bridge the gap between multidisciplinary users of diverse tools; exchangeable geometric models are thus a particularly pressing need. Success in developing exchangeable representations of performance, geometry, and process requirements is a prerequisite for their use in design tools and by practitioners in the allied domains of process equipment design and shop floor planning and operations, as well as by designers in other companies or in other technical domains.

• *Tools that facilitate understanding the relationship between cost and de-*

> ### BOX 3.6 Limitations of Tolerance Analysis
>
> The inherent variability in the shape of manufactured parts imposes major limitations on the performance of products assembled from those parts. To reduce the impact of such limitations, designers traditionally impose tolerances on these parts. However, too little is known at present about the relationships between unwanted variations in shape and product performance to permit rational assignment of tolerances except in a few special cases such as optical systems and journal bearings. In most cases, tolerances are assigned on the basis of experience or the capability of available manufacturing systems. Two fundamental tolerance problems are tolerance analysis (i.e., identifying which dimensional variations contribute most to a final error) and tolerance allocation (i.e., choosing the best way to distribute inevitable tolerances among several potential contributing sources during design so that the desired level of final error is achieved at minimum cost). Today, most tolerance analyses are simple one-dimensional fit studies. More complex multidimensional studies are done less often and employ Monte Carlo methods. These methods are difficult to use because there is no automated way to apply them to a computer model of the geometry. They also suffer from combinatorial and scale-up barriers. Research is needed to extend engineering models and then link them to geometry and geometric variations.

sign choices. Design tools for electronic devices typically do not provide information about fabrication cost, because the cost of fabricating a given chip is approximately independent of what is put on it.[8] But this is not true for mechanical design, in which design choices may have a significant impact on the cost of making the final product. Designers need to be able to keep such high-level trade-offs constantly in mind through the use of better design tools. Many papers have been written on this subject, and some such tools exist (e.g., Hewlett-Packard's Sheet Metal Design Advisor).

More generally, product data models will have to include information that goes beyond simple geometry. These models should also relate high-level function, assembly processes, possible interactions between product and user, the product's structure and geometry, a schema for describing component parts, and the product's behavior, manufacture, maintenance, and ultimate disposal or recycling—in principle, any information that could describe a product. Moreover, the information contained in a product data model may represent a single instance of the product, a set of possible instances, or even a set of possible descriptions.

---

[8] One qualification is necessary. While chip fabrication is generally pattern-insensitive, the yield (i.e., the fraction of chips that are usable after the fabrication process is complete) is a direct function of the feature sizes and material gaps chosen: wider lines and wider gaps between lines generally result in a higher yield, which in turn reduces the unit costs of chips that may be sold.

BOX 3.7 Using Design as a Strategic Tool

Nippondenso Co. Ltd. (NDCL) manufactures automotive components. Its order mix is highly variable and unpredictable, and thus a primary challenge for NDCL has been to develop a factory environment in which it can produce a near-arbitrary mix of product variants. This type of flexible manufacturing has often been attacked as a problem involving factory floor operations, but NDCL has defined and solved it primarily as a problem of design.

In particular, NDCL's design approach is based on what can be called a combinatoric method (Figure 3.1). Product variety is created by selecting among several versions of each part in a product. Each product is designed so that the physical and functional interfaces between parts are the same for all versions of each part. The result is that any combination of versions of parts can be assembled into a working unit. The interfaces between the parts and the assembly equipment are similarly standardized so that differences between versions of parts are transparent to the equipment.

NDCL achieves its flexibility goals by using fairly ordinary parts and then employing unusual logistical methods to assemble them into different variants. The schedule for fabrication of parts is not based on the details of what kinds of items are ordered, but instead follows broad statistical patterns of orders. It is the assembly process that addresses the detailed stream of orders. Because assembly is so much faster than fabrication and because NDCL's assembly machines can be switched from model to model so quickly, model mix can be addressed much more easily and economically during assembly than during fabrication.

Close coordination of top management objectives, product design, and production technology is required to carry out this approach. As a result, it can be said that NDCL has taken concurrent engineering well beyond the goal of improving fabrication or assembly. Instead, NDCL has learned how to use design to achieve the essentially strategic goal of meeting the demands of its customers.

Thus, in addition to serving the traditional role of providing neutral data formats for computer-aided application tools, models must support a host of activities for managing and exchanging information.

Research for Process Description

Tools for process description have application to the design and operation of factories; in addition, they are the basis for experiments with and evaluation of control and organizational changes before actual systems are installed. Process descriptions will also be used to enhance product design, so that by simulation the best process can be matched to the product design (and vice versa) for maximum economic advantage (or to satisfy whatever criteria—such as quality or time to delivery—are important for the particular case). Box 3.7 describes a rich and productive interaction among process design, product design, and what happens on the shop floor.

The primary need in process description is formalization, which is necessary

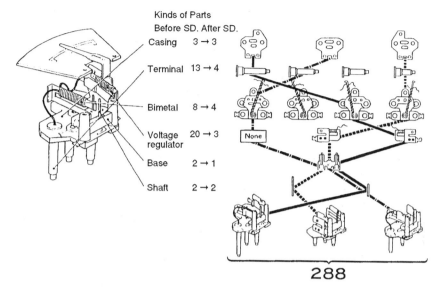

FIGURE 3.1 A panel meter (left) and the combinatoric strategy (right). Each zigzag line on the right represents a valid type of meter, which is assembled by following the path from top to bottom. "SD" stands for standardized design, an effort that reduced the number of variants of each part as shown. The production rate is 32,000 per shift. A "catalog" of only 16 parts is sufficient to support production of 288 different kinds of meter. If all the 288 possible paths at the right were drawn, one would see that each part is a member of many possible types of meters. Thus to first order most parts will be used regardless of the pattern of the order stream, so that there is little inventory risk in making the 16 kinds of parts. If each different meter type were created by a few parts that were special to that type, a shift in the order pattern would require a large and awkward shift in the schedules for fabricating parts, which could not be accomplished as quickly as the switching of an assembly machine. Also, feeding the hundreds of different kinds of parts needed to support so many varieties of meter would be very awkward. These are some of the reasons that variety can be achieved more easily during assembly than during fabrication. Courtesy of Nippondenso.

for representing processes in sufficient detail and with enough specificity to make the process description adequately complete and unambiguous.[9] Such formalisms allow designers to describe, enforce, and simulate processes, including fac-

---

[9] Of course, at the root of a good process description is a good scientific and engineering understanding of the specific process involved—the best tools to formalize process description are not helpful if the knowledge base they are used to formalize is shaky and uncertain. Indeed, many fabrication processes in use today are incompletely understood and only partially characterized. Still, as important and crucial as such understanding may be, a research agenda for process characterization is outside the scope of this report.

tory processes (involving both machines and people), design activities, and decision processes.

Research needs in the area of process description include:

• *A language for expressing process descriptions that facilitates checking for correctness and completeness and the capability to express not only nominal process behavior but also variant behavior.* Such a language must also be translatable across technical domains.

• *Process model representation schemes (both aggregate and detailed) and on-line data collection.* Data included in such schemes should describe logistic, fabrication, material-handling, inspection, assembly, and test processes and should give information on the characteristics, capabilities, and costs of various production and assembly methods. Data should be captured as a product is being produced so that the process can be improved.

• *Specific process models that reflect all relevant spatial and temporal transformations.* Such models are critical for the local control and planning of manufacturing operations and will draw on knowledge about the kinematic capabilities of individual pieces of equipment and other process limitations, processing capabilities of the equipment, and tool and fixturing capabilities associated with the equipment. Ultimately, these models should contain the detail necessary for dynamic control of the individual operations as well as the information required to simulate the operation of the manufacturing system, indicate the effects of perturbing the operational parameters as well as the effects of complex interactions among processes, and be generalizable across a wide range of production environments by appropriate parameterization.

• *Algorithms and tools to solve process problems.* For example, one such problem is the determination of efficient assembly sequences; an inappropriate assembly sequence may result in the need for a tool to reorient the item being assembled many more times than necessary, thus increasing the time needed for assembly and the likelihood of breakage. Such factors are to a considerable degree irrelevant to the design of the product itself but have a significant influence on the cost of making the product. Other important process problems include the determination of efficient equipment layouts on a factory floor, equipment selection (matching equipment capability to process needs), make-buy decisions, and determination of how best to cut and shape materials to minimize waste.

• *Dynamic models for describing resources available over time to the manu-*

facturing system.[10] Such models will be used in both the design and the operational control of manufacturing systems. Common representations and descriptions of resources are necessary to enable development of transferable (from planning to analysis to control) models and analysis. Despite the importance of resource management, little research and effort have been devoted to creating generic representations of resources.[11] As a result, specific resource characteristics must be recreated each time a modeling activity is undertaken. Resources include those related to fabrication (e.g., tooling, machines, available controller features) and their interconnection, as well as system resources such as corporate information or knowledge and information such as company or external standards.

Research for Tools to Support Integrated Product and Process Design

Computer tools that directly aid the management of the IPPD process itself would be helpful to managers. Today, such tools are limited primarily to communication aids or "groupware" for helping people post notices and share information. Means of describing and managing the design process need to be developed. Few tools exist for creating, monitoring, and guiding the design process itself, except for familiar project management tools like PERT. (PERT is largely a schedule and resource management tool.) Scheduling aids beyond PERT are required to help in determining effective task sequences, setting up information flows, establishing schedules and milestones, identifying people, assigning work to them, routing information to them, and linking them to colleagues elsewhere. Existing tools do not help to identify information flows or facilitate them.

A single design decision may have several simultaneous impacts, some of which may be beneficial and others adverse. The design environment should support techniques to express comparisons and trade-offs vividly so that a designer can assess the impact of a wide range of design decisions on a product's cost, time to produce, or quality. Tools are needed that focus on the identification

---

[10] See MSB (1988), pp. 11-13. It recommends research on a number of specific resource management modeling methods, for example, modeling methods based on knowledge-based systems, object-oriented systems, and Petri nets; methods that are sufficiently fast and efficient that resource problems are tractable while plants are being designed and built, as well as being operated; methods for correcting models, based on comparisons of predicted and measured performance; and many other methods.

[11] A notable exception is IDEFx, an evolving language developed for the Air Force in the Integrated Information for Concurrent Engineering program that does identify information flows within the process and is used to support the reengineering of business processes through process modeling and simulation. The first version, IDEF0, was developed under the Integrated Computer-Aided Manufacturing program and was used primarily for modeling of individual activities within processes; IDEF0 is today a Federal Information Processing Standard. See Moore (1994), p. 49.

of design trade-offs between cost, performance, and reliability; alternate space allocations; functional decomposition; subassembly definition; three-dimensional geometric reasoning; and make-or-buy decisions.

To produce such tools, research is needed on decision tools that draw on the product-process data model and performance simulations of the product being designed, as well as process models and various data on costs and tolerances of different processes. These decision tools will sort data and models on the basis of criteria supplied by the designer to aid in making comparisons between alternate designs and processes.

In addition, improvements to simulation and rapid prototyping tools should be brought to bear on the problems of viewing physical parts, "visualizing" complex relationships (certainly between physical parts, but also between more abstract relationships such as design requirements and their costs), and presenting design alternatives to customers and to designers. This category also includes methods or tools to handle groups of parts such as assemblies, subsystems, product families, made-to-order configurations, and selected combinations of parts that create different product models by virtue of which parts are selected. (Box 3.7 describes such an application.)

Research is needed on design methods and computer tools adapted for the design and creation of groups of parts or systems, in addition to individual parts. Such methods and tools will enable the designer to divide a product into subassemblies, design optimal in-process test strategies during assembly, and identify assembly sequences that minimize cost, tolerance errors, material handling, and part damage during assembly. Also, the methods and tools will include standard design modules and methods that facilitate optimization of part or all of a product for cost and quality. Such optimization requires a deep understanding of specific components and features of the product.[12]

Finally, it would be highly desirable to have tools that would allow product design and process design to proceed more in parallel. These tools would enable product designers to work with some degree of incomplete information about the process designer's work, and vice versa. Successful development of such tools would contribute greatly to the reduction of needed design time.

Research Areas Not Specific to Manufacturing

Geometric Reasoning

A generic intellectual activity required by mechanical design is geometric reasoning. A major difference between VLSI design and mechanical design is

---

[12] Because such understanding is often proprietary to component suppliers, research in this and similar areas of collaborative design must consider nontechnical issues such as intellectual property rights and intercompany data exchange.

the degree to which three-dimensional geometric reasoning is fundamental to even the simplest mechanical designs; VLSI design is based on two-dimensional (2-D) analysis (or at worst, 2 1/2 dimensional analysis—the use of stacked 2-D layers). Three factors make three-dimensional (3-D) reasoning more complex: 3-D design is more likely to involve moving parts or flows; 3-D design may involve interconnections and tolerance relationships between 2-D domains; and 3-D designs can be visualized only in cross sections, perspective, and exploded views. Research is needed on building robust geometric modelers. Many of the "boundary representation" modelers in use today are not robust: interactions between the algorithms they use and finite-precision arithmetic offered by the computer result in certain modeling operations that yield incorrect results. Truly robust algorithms to correct such problems remain a challenging research problem. Ultimately, the design environment should support improved visualization tools or other design aids that will help make geometric reasoning faster and efficiently achievable by a broader range of people. In addition, tools that undertake geometric reasoning automatically (i.e., without relying on a skilled human) may be able to replace human designers for certain purposes.

Knowledge and Information Management

Basic to design are many issues of data management and of data themselves. The future design environment will include a number of data management methods and tools. For example, the design environment will have to handle data legacy issues, such as converting data from one CAD system to another and preserving old data for decades or more so that they can still be read, edited, and processed. Today, such data are either lost, kept on paper, or accessed in a limited way by old hardware kept on hand for the purpose.[13]

A second data management issue is that the design environment will have to include ways to capture corporate memory and knowledge so that successors of current designers can tell what knowledge was used, what competitive methods were used, what errors were made, and on what factors success was based. The ability to record design history and rationale is of particular importance. Every design is a historical web of decisions that grow out of each other and depend on each other. Revision, whether for correcting an error, absorbing a new outside circumstance, or improving manufacturability, requires unraveling the web to a certain degree. The design selected, the web of decisions leading to it, and "roads not taken" indicate the corporate state of belief at the time the decisions were made and thus form a historic context.

---

[13] Siewiorek (1992) posed the following research questions that must be resolved before concurrent design and rapid prototyping become integrated into industrial practice: How can design and manufacturing information be reused in future products? and, How can the compatibility of new incremental information with all the previously acquired information be ensured?

SPECIFIC RESEARCH QUESTIONS

A variety of specific research questions are motivated by discussions earlier in this chapter. In accordance with the charge of the study, the focus of these research questions is largely information technology and associated disciplines such as mathematics. If the underyling science and engineering aspects of manufacturing are well understood, IT can help immeasurably in exploiting such knowledge and information, but by itself IT is not a substitute for that knowledge.

- How should the data contained in a product data model be organized to accommodate their huge size and complexity and the many disciplines that need access to them?

- How much of that information is physical and how much is "relational"? How much can be captured in traditional geometry and how much is nongeometric or not focused on one item but shared or spread among many items, or even not attached to specific items?

- To what extent can a product description for mechanical items be converted *automatically* into a production plan, that is, a sequence of fabrication steps that transform raw materials into a final product?

- Can a general process description language be developed that would be both man- and machine-intelligible, permitting processes to be described more precisely than is possible now?

- How can CAD tools be used at different stages of the design process? Can high-order abstractions be used as a starting point for encapsulating product and process facts and knowledge, geometry, requirements, tolerances, and other product characteristics, providing a link between product function, geometry, and processes? Can "meta-features" be defined that will encapsulate groups of features, creating a feature hierarchy? If not, how can the scale of real designs be encompassed using features or any other means of capturing and combining detailed design data, geometry, and intent?

- Features are often very process-dependent, and the fabrication of parts may require the application of multiple processes, each of which uses a different set of features to address the same part. Attaching features to descriptions of parts may prove impractical due to the differing feature sets for the same part. Feature-based product description may prove unproductive. In this event, to what extent is it feasible to design process models that interact with separate product models to generate appropriate product models?

• How can the human interfaces of CAD systems be improved so that more complex shapes, assemblies, and other multidimensional problems can be handled more easily?

• What new process analysis tools can be developed, especially to handle complex problems like assembly, model mix manufacture, and tolerances?

• How can the reliability of new product or process designs be better predicted, including trade-offs between cost and reliability?

• What new languages or data structures can be developed to better describe product requirements, such as performance, reliability, shapes, interconnections, interfaces, and tolerances?

• Similarly, what new languages or data structures can be developed to better describe process requirements, such as performance, cost, reliability, ease of diagnosis and repair, material handling, ease of use, and ease of modification?

• What new data logging and correlation methods can be devised that would help the process of continuous improvement, such as finding multiple occurrences of the same type of machine failure or deducing what is the best sequence for testing a broken system to diagnose its problems?

4

Shop Floor Production

INTRODUCTION

Several generic tasks characterize production, the process through which parts and materials are transformed into final products. These tasks include, among others, the receipt and acknowledgment of orders, the acquisition of materials, the performance of shop floor operations, and the generation of information needed to support continuous improvement. Together, these tasks (when properly done) constitute a qualified production process. Qualifying a production process is a demanding and important task that requires people trained and physically qualified for a given job, machines and process instruments that can be guaranteed to operate within specifications, production capacity that can match the order demand, and the availability of production capacity in the desired time frames.

The information-processing view of a production facility is in essence the same as that for an individual work cell within the facility. Both factories and work cells process orders and turn out products. For a factory, the order usually comes from a customer outside the factory; for a work cell, the order comes from inside the factory. For a factory, the product delivered is a final product that can be sold to an external customer; for a work cell, the product delivered is a partially finished product that goes on to the next work cell, which regards it as a part or a raw material for that next cell. The demands made by factories on suppliers for components are the same as the demands made by an individual process to be carried out by a work cell.

This chapter focuses on those aspects of production related to the scheduling of specific factory activities, control of the activities and the operation of machines on the shop floor, and mechanisms for providing rapid feedback that becomes the basis for near-real-time adjustments in various production activities. Table 4.1 summarizes information technology (IT)-related research needed to advance shop floor and production systems.

SCHEDULING FACTORY ACTIVITIES

Centralized Control

The dominant issues in production planning today are achieving major reductions in manufacturing lead time and major improvements in honoring promised completion times. Better scheduling and planning might well help to reduce the total time required for converting raw materials and parts to finished products.

Today's dominant scheduling paradigms are based on material requirements planning (MRP) and manufacturing resources planning (MRP-II), although other approaches are used from time to time. MRP and MRP-II were developed in the 1970s and 1980s. Originally developed to handle planning purchases of parts required for the products to be manufactured, MRP assumes a plant of infinite capacity. MRP-II goes beyond MRP to take into account inventory, labor, actual machine availability and capacity, routing capabilities, shipping, and purchasing. MRP-II generates a master production schedule that serves as the driver or trigger of shop floor activities.

By design, the characteristic time scale of scheduling (in MRP jargon, the "bucket") based on MRP and MRP-II is days or weeks, perhaps even months; depending on the implementation, MRP and MRP-II may or may not be the basis for "to-the-minute" or "work-to" timetables for equipment use and material flows. Under any circumstances, however, arranging for the moment-to-moment control of factory operations is the job of the shift supervisor. For the shift supervisor, situational awareness is at a premium. He or she must draw on information provided by workers on the previous shift (reflecting matters to which he or she must attend on this shift) and sensors and worker reports of what is happening on the shop floor during this shift (e.g., the condition of various tools, how various processes are operating, who has reported for work, what materials are available), as well as directives from senior management concerning overall objectives. From these sources, the shift supervisor must develop a local plan of operation for the next 8 hours.

In principle, the real-time factory controllers of today operate by periodically receiving a list of jobs that must be completed if the MRP-based scheduling plan is to be followed. For each production job, a process plan is retrieved from a database. Such a plan involves the identification of routings of particular work in progress to specific process cells, scheduling of all activities at both the cell and workstation levels, coordinating those activities across the equipment, monitor-

TABLE 4.1 Research to Advance Shop Floor and Production Systems

Subject Area	Example of Research Needed
Equipment controllers	Architecture and technology for shop floor equipment and data interfaces
	Open architecture for control systems
	Appropriate operating systems, languages, data structures, and knowledge bases
	Human-machine interfaces to permit people to interact effectively in a modern manufacturing environment
	Design for repairability and the ability to work around equipment crashes, including diagnostic software
	Better real-time control
	Wireless communication
Sensors	Standardized interface connections
	Manufacturing control architecture
Dynamic (real-time) scheduling	Dynamic shop floor models with high-speed recomputing time and the ability to handle numerous variables
	Real-time planning and scheduling tools for the flexible factory and the distributed factory
	Techniques to ensure graceful degradation of production operations in the event of local problems
	Tools to facilitate situation assessment and scheduling by the factory manager and operations team
	Multilevel understanding of large-scale systems
	Means for identifying the relevant measures and quantifying the relative performance of alternative systems
	Tools to support the brokering of priorities and obligations among cooperating entities, based on minimizing overall transportation, material handling, inventory, capital, and labor costs
Intelligent routing systems	Identification of appropriate interfaces among product design, product engineering, manufacturing engineering, and factory floor procedures as they emerge in computer-augmented work groups

TABLE 4.1 Continued

Subject Area	Example of Research Needed
Intelligent routing systems (*continued*)	Demonstration of the resilience of intelligent routing systems with respect to the vagaries of factory conditions
Smart parts (automatic routing)	Practical open standards for recording and communicating data among parts, assemblies, subsystems, and their network of makers and maintainers
	Mechanisms for cost-effective embedding of information and for ensuring access throughout the life of a part
Modeling of manufacturing systems	Ways to efficiently manage large amounts of related data spread over many machines and locations
	Ways of specifying complex data relationships
	Ways of ensuring interoperability of design process tools
	Better ways of encoding and decoding data
	Improved data retrieval methods, including human interfaces
Rapidly reconfigurable production systems	Development of agility in the face of rapid change in a number of important product or process variables
	Investigation of the feasibility of developing a reasonably universal product configuration language and methodology
	Assessment of the feasibility of using programming languages to represent manufacturing operations in the same sense that design languages represent designs
	Development of systems that simulate the operation of a given manufacturing configuration under a variety of conditions to optimize configuration
Resource description models	Development of schemata and models to represent manufacturing resources and their interconnection
Knowledge bases for new process methods	Development of a robust and flexible system that can model efficiently nearly any process that may be developed

ing activities, and job tracking. The real-time control system then determines the sequence of activities necessary to execute that process plan. Once determined, the sequence of activities is passed down a control hierarchy that is often organized around equipment (e.g., numerical control machining centers, robot systems, automated ground vehicles, or any other computer-controlled manufacturing equipment and related tooling resources), workstations of interrelated equipment, and cells of interrelated workstations.

In practice, the activities on the shop floor in most MRP or MRP-II installations are orchestrated by an informal, manual, chaotic process that tries to adhere to the assumptions on which a given MRP scheduling plan is based. Such plans are difficult to change when unforeseen contingencies occur (e.g., delays, material defects, unexpected machine downtime). Thus, in an attempt to avoid disrupting the schedule, factory managers expend considerable effort in solving real-time shop floor problems, for example, by arranging for priority delivery of parts at added expense. Nevertheless, considerable replanning and rescheduling are necessary when managers are unable to fix these problems, a not-infrequent occurrence in a manufacturing enterprise.

Effective real-time control of a factory depends on scheduling that remains close to optimal (relative to some performance measure(s)) over a realistic range of uncertainties and over time. With such scheduling, priorities from moment to moment can be balanced against circumstances prevailing in the plant and in the manufacturer's supply chain. Such circumstances might include sudden changes in conditions generated by drifting machine capability, material shortages, worker absenteeism, unplanned downtime, tool breakage, and the arrival of unexpected rush items. Today, managing such sudden changes often requires costly crisis intervention. An effective real-time schedule must be flexible enough to absorb minor perturbations in the system and require many fewer "reschedulings" when something goes wrong. Real-time dynamic scheduling with this kind of flexibility would be able to reflect the external priorities of the manufacturing enterprise. Dynamic scheduling would increase the overall utilization of material, labor, and equipment by improving the whole system and modifying the operations at each cell, thus reaching levels unattainable with the best of present methods; for example, dynamic schedulers would account for interactions among receiving, material handling, and cell schedules and propagation of changes.

A dynamic scheduler would continually track the status of jobs, cells, tooling, and resource availability. Through network communications, each cell would have access to information pertinent to the shop floor. What should be done next by any particular cell at any particular moment would thus be determinable at that cell and would be based on current conditions throughout the factory.

A follow-on to MRP/MRP-II technology would provide tools for production planning, scheduling, and control with:

- The ability to build and maintain high-quality finite capacity schedules (both aggregate and resource-based) that realistically account for the major constraints and preferences characterizing the production environment at hand;

• Integration of adaptive planning, scheduling, and control capabilities, thus making it possible to quickly revise a production plan or schedule to account for a wide range of contingencies and making it possible to drive automated equipment in real time and to make trade-offs of local goals in favor of higher-level ones;

• The ability to effectively integrate improvements in a wide range of production planning and scheduling decisions within a given manufacturing site as well as across the supply chain;

• Support for powerful interactive capabilities making it possible for the user to incrementally manipulate the production schedule at multiple levels of detail and to explore "what if" scenarios in overall factory planning as well as scheduling on the shop floor, and helping the user to identify possible inefficiencies in the production schedule and determine ways to correct these inefficiencies (e.g., adding overtime, shifting personnel around, buying new equipment, modifying raw material reordering policies, and so on);

• Reconfigurable and/or reusable software, making it easy to customize these techniques for a wide variety of manufacturing environments;

• Support for integration with other important functionalities (e.g., process planning, factory layout, accounting, preventive maintenance, and so on);

• Capabilities to deal with the stochastic and sometimes reentrant nature of shop floor events. Within the MRP/MRP-II paradigm, inputs are deterministic estimates of the times for various shop floor operations that are obtained by time and motion studies. However, in reality, the timing of events is probablistic rather than deterministic, and the distributions of service times and of interarrival times, far from being deterministic, look like nonstationary exponential probability density functions with long tails.[1] Manufacturing operations may also be reentrant, in the sense that the same equipment is often used for performing several different steps at different times during a production process (e.g., photolithography in semiconductor manufacturing). Such nonlinearities introduce added difficulties for the development of new scheduling paradigms; and

• Backwards compatibility with MRP/MRP-II technology. While technologists often believe that a clean break with current technology will enable newer technologies to avoid their most basic problems, as a matter of practicality, few manufacturers will be willing to abandon familiar though perhaps flawed systems in favor of new technologies not proven in the factory. Whether this means that the follow-on to MRP/MRP-II will be an entirely new paradigm for

---

[1] Personal communication: Charles Hoover, professor of electrical and mechanical engineering, director of the Manufacturing Engineering Program at Polytechnic University, Brooklyn, New York; July 1994.

technology with translators that bridge the gap to legacy systems or an evolutionary improvement to MRP-II is an open question.

To support effective scheduling, research is needed in the following areas:

• *Dynamic shop floor models with fast run times and the ability to handle
numerous variables.* Factors that affect production scheduling include the
following:

> • Are the appropriate resources available at the right time? (Such re-
> sources include people, tooling, jigs, fixtures, or other special product or
> process needs.)
> • Is the equipment up and running?
> • What is the next scheduled downtime? Will people be there to handle
> this?
> • Is the "downstream" process ready to handle the output?
> • What is the status of any inventory buffers?
> • Can a particular lot be fabricated using factory resources other than the
> one for which it is planned? If so, how?

All these and other variables have to be entered into the factory model and
scheduling system in order for the results to have any benefit. The model
could include agents that would take requests from customers (i.e., other
agents that are "fed" information provided by these agents), decompose
those requests into information requirements, and then request that information from its suppliers (i.e., still other agents that feed information to these
agents). As suppliers respond with estimates of their capability, the agents in
question make realistic forecasts of their ability to respond. Once the re-
quested information is in hand, the agents pass status reports to customers.

• *Real-time planning and scheduling tools for the flexible factory and the
distributed factory.* Such tools would also provide capabilities for integrating scheduling and control reactively (e.g., they would make use of adaptive
scheduling techniques based on the severity of the contingency at hand and
the time available to repair the schedule and/or scheduling techniques that
could exploit windows of opportunity that occur fortuitously) and for propagating of schedule changes throughout the system.

• *Techniques to ensure graceful degradation in the event of local problems.*
Many production operations today are brittle, in that a problem in one crucial
location can halt an entire production line. More desirable would be a real-
time scheduler that could detect cell problems as they occur and reroute work
flows around them while the cell in question was being repaired. If such
dynamic rerouting were possible, factory output would be damaged only to

the extent that one cell's contribution to the overall process was no longer available, rather than producing the functional equivalent of losing the contribution of many cells to overall output. Dynamic rerouting would also enable the production of varying items. Research in this area would address diagnosis of plan failures, plan repair, and planner modification, as well as analysis of material handling.

• *Tools to facilitate situation assessment and scheduling by the factory manager and operations team.* For the immediate future, management of shop floor operations will be under direct human supervision. Human decisions makers will need such tools if they are to perform their jobs effectively. Tools in this area should also provide techniques for analyzing interactive planning and scheduling. Better techniques would help to develop tools to monitor and analyze the operation of an enterprise and identify bottlenecks and opportunities. One important aspect of this problem is avoiding suboptimization by understanding the impact of alternative goals at each workstation and downstream consequences of changes made. In addition, since such tools are likely to be able to produce scheduling-relevant information in voluminous quantities, good visualization and other human-computer interfaces will be necessary as well.

• *Tools to support the brokering of priorities and obligations among cooperating entities,* based on minimizing overall transportation, material handling, inventory, capital, and labor costs. Such tools must also integrate multiple dimensions in which managers must make decisions, including release decisions (e.g., when to release jobs and orders to the shop floor), reordering decisions (e.g., when to reorder raw materials and/or components, how much, and from whom), sequencing and batching decisions (e.g., grouping orders for similar parts to reduce setup costs), safety stock and safety lead-time decisions (e.g., increasing lead times as a hedge against various sources of uncertainty), overtime or order-promising decisions (e.g., promising delivery dates to potential customers), coordination decisions (e.g., using distributed scheduling and control within a given plant as well as across the supply chain), and material-handling decisions (e.g., when to add capacity or when to change the strategy of the material-handling system).

• *Development of new techniques for optimizing production planning and scheduling.* Traditionally closer to operations research (OR), several successful optimization techniques are better characterized as a mix of techniques drawn from artificial intelligence and OR.[2] Learning can conceivably

---

[2] For example, certain scheduling techniques developed at Carnegie Mellon University are based on constraint satisfaction techniques developed in the artificial intelligence literature and also combine various OR-like optimization subroutines.

be used at all levels to enhance the performance of these tools (e.g. to enhance the performance of schedule-optimization techniques, help determine strategies for selecting among multiple reactive control policies, learn important user preferences to facilitate interactive planning and scheduling, learn to identify specific types of inefficiencies in the production schedule, and learn effective coordination strategies with suppliers and subcontractors). A related research problem involves the development of optimization techniques that lend themselves to efficient implementation on parallel hardware.

Decentralized Control

An alternative to top-down scheduling of production activities is a decentralized organizational structure in which activities are initiated on the shop floor itself, without regard for a master factory scheduler. Since it is not known with confidence whether a top-down or a bottom-up approach to initiating and performing production activities is more likely to result in success, it is reasonable to investigate both approaches. At least two aspects of decentralized control seem worth exploring, autonomous agents and work and logistics flow.

Autonomous Agents

The use of autonomous agents may offer some potential as a means to handle complex dynamic environments. Implemented as software objects or collections of objects (perhaps representing physical robotic agents),[3] they may provide solutions for manufacturing problems in the areas of planning, monitoring, and control. Agents may be able to take responsibility for shutting down a machine, starting up a program, or sending a message to another agent. Because such agents automate control and interaction functions, they can eliminate activities otherwise performed by people and allow for simpler organizational structures, which in turn can simplify requirements for software development and maintenance.

If autonomous agents are to be successful, better architectures for manufacturing systems involving distributed intelligence will need to be developed. Agents should model real-world behaviors, enable encapsulation, promote flexible distribution of control versus centralized control, capture the inherently

---

[3] The committee recognizes that the precise definition of an agent is a topic of debate in the community. (A significant amount of time was spent engaged in this debate in the AAAI Special Interest Group for Manufacturing (SIGMAN) workshops of 1992 and 1993, and the IJCAI SIGMAN workshop of 1993.) Another definition of an agent is that it is an actor with certain cognitive characteristics, such as motivation and intent.

nondeterministic concurrent nature of the environment, and model the needed level of knowledge or behavioral detail at any instance in time.

A critical consideration for sensible agent behavior is level of autonomy. The practical deployment of agents in manufacturing requires that they behave sensibly, incorporating an understanding of both global and local goals. The issue of agent autonomy is very relevant to planning, monitoring, and controlling agent behaviors. The purpose of planning behaviors is to develop and plan steps of action. Monitoring behaviors involves evaluating the environment external to agents. Controlling behaviors encompasses modifying behavior or data. The extent of autonomy can range from local autonomy, which requires that agents be capable of initiating their own thread of execution, to consensus (negotiation) autonomy, with each agent adhering to a set of rules or constraints governing how consensus is achieved and ultimately how agents behave, to command-driven autonomy, with agents restricted to simply executing some given request (command) without the benefit of queries or "fact finding" about the state of the global world (state of agents external to a given agent).

Research is needed to develop tools to find information (such as autonomous agents), tools to distribute information, and stable sets of rules for interacting agents (rules for agent behavior). With regard to development of autonomous agents, the following research topics appear important:

• *Architectures supporting dynamic distribution of intelligence among autonomous agents* for planning, monitoring, and control as well as dynamic levels of autonomy.

• *System stability.* For sensible decision-making capabilities, autonomous agents must understand how to resolve conflicting goals. Instabilities may arise when multiple agents interact, although collections of agents may also demonstrate emergent characteristics, that is, modes of stable (and desirable) behavior that arise from the complexity of their interaction. Strategies for resolving conflicting goals may need to be developed.

• *Better representations of spatial and temporal dimensions relevant to shop floor operations,* rather than simple declarative knowledge.

• *Representations supporting multiple perspectives of data and knowledge.* Simulation, analysis, and design tools need to use product information. Yet each of these applications of the data has a different perspective on the product data. These data must be consistent and persistent, and translators often fall short.

To date, no autonomous agent built has demonstrated sophisticated behavior. Even a task such as searching the Internet for the e-mail address of a particular individual remains out of reach, although many researchers argue that an agent with this particular kind of capability will soon be available. If the potential of agent technology is to be broadly believed, a convincing demonstration must be created.[4]

Work and Logistics Flow

A complementary approach to achieving decentralized control is the use of an intelligent routing system that can route a partially finished product to the next available manufacturing cell or station capable of performing the work needed in the next step. In such a system, each manufacturing cell knows the operations that it has been certified to perform and bids on work to signal its availability when it is free. Cells communicate their status to intelligent parts carriers through a communications network. Work flow is thus controlled in a bottom-up manner.

One approach to an intelligent routing system presents acceptable methods (or routes), including one determined to be the best according to specified criteria (e.g., minimum cost, maximum speed, and/or maximum quality). All acceptable methods become functional alternate routings. Whenever an engineering change occurs, the routing is downloaded to all cells that have been authorized by the engineering change to bid for work on the part.

Another major aspect of shop floor management is coordinating the flow of various components to appropriate locations on the shop floor. Today, the transport of parts to these locations is performed by intelligent parts carriers. These carriers are in essence automatically guided vehicles that carry parts to any available workstation or to the preferred workstation (selected, for example, on the basis of its having the shortest queue). The carriers expedite the progress of parts through the plant.

Such carriers are most useful in a single facility. However, when products require production that crosses building or enterprise borders, "smart parts" will carry the information necessary to supply production instructions and history. (A simple illustration of a smart part is one that signals its location within a factory as it is moved about, perhaps through the use of bar codes or attached radio transmitters.) As manufacturing operations are carried out, instructions embedded in the part can be deleted or marked to indicate completion. Smart parts can also monitor and indicate their own performance. Thus maintenance requests can be triggered if the part senses its own performance to be substandard. Planned maintenance regimens may also be recorded within the part, automatically triggering requests for maintenance.

To realize intelligent routing systems, research is needed to identify appro-

---

[4] Even if no single agent will be able to demonstrate sophisticated behavior, the possibility remains that a collection of agents may be able to do so.

priate interfaces among product design, product engineering, manufacturing engineering, and factory floor procedures carried out in computer-augmented work groups, including human-computer interfaces. The immediate research focus should be on demonstrating the resilience of the intelligent routing system with respect to the vagaries of factory conditions.

For smart parts, research is needed to identify practical, open standards for recording and communicating data among parts, assemblies, subsystems, production equipment, and maintenance equipment. Research is also needed into cost-effective mechanisms for embedding the information and for ensuring access throughout the life of the part.

CONTROLLING INDIVIDUAL FACTORY ACTIVITIES

Production (e.g., material modification, assembly, testing, recycling, or information processing) takes place in a factory. The factory environment is inherently messy and highly unpredictable: dust obscures lenses, objects fall in the wrong place, and metal-working lathes jam. In this uncertain environment, a factory control system determines the operational production efficiencies that influence greatly how responsive a system can be to customer orders and how much a product costs. Since all business, engineering, and integration activities come to fruition with the production of an item, the factory control system must therefore "hook to" all of the other business and engineering activities as well as to those on the shop floor itself.

Information technology has the potential of taming the future production environment by making it more flexible, efficient, and responsive. For example, information technology enables the increased automation of traditional production techniques, as well as new techniques such as stereolithography and material deposition. In addition, many of the individual technologies and components of unattended machine tools and other fabrication and assembly equipment already exist as experimental devices in the laboratory or as commercial products, and in Japan, several "lights-out" factories (factories that operate with limited human involvement) have existed for several years.

However, a fully integrated computer-based factory control system that is flexible and robust is still very much a goal rather than a reality. Major unsolved research problems remain, including those discussed later in this chapter. Moreover, vendors and technology suppliers have chosen in many cases not to build on previous work, instead opting to develop and use their own proprietary standards and interfaces, while in other instances, companies have developed their own standards. Today, even so "minor" a characteristic as the amount of data that must be passed between two points in a factory can slow down control systems to the point that they are no longer functional. In some instances, even products by the same vendor are incompatible with each other.

Equipment Controllers

The purpose of an equipment controller is to effect a timely change in the state of the piece of equipment being controlled. State changes effected too early or too late can cause system instabilities and even failure. Controllers are used in many aspects of shop floor operations; factories employ mill drive controllers, cutter controllers, controllers for variable-speed pumps, crane controllers, controllers for large AC/DC rectifiers, controllers for material-handling systems, and automated guided vehicle controllers. For simplicity, the discussion below focuses on tool controllers, although many of the issues discussed also arise for controllers of other types of equipment.

A tool controller translates human directions into specific actions taken by the material-working end of the equipment. The simplest controller is the classical proportional integral-derivative (PID) controller, which uses the basic time constants of analog electrical circuits to control timing. In recent years, these analog controllers have been supplanted by computer-based digital controllers, because digital controllers are not subject to drift or other calibration inaccuracies and because the simple analog controllers are not adequate to drive more complex assemblies of tools.

Today's computer numerical controllers (CNCs) are generally adequate for servo control of motion axes and control of specific discrete devices, but they do not adhere to industry-wide standards for intercommunication. As a result their programming flexibility and ability to communicate with external computers and devices are limited:

• Standard configurations cannot accommodate nonmachining devices such as work-holding accessories, force sensors, vision sensors, and other subsidiary devices.

• CNC design concepts are excessively conservative, especially with respect to their hardware interfaces. For example, current CNC communications are effected mainly through slow serial lines (RS-232 and other similar interfaces), or through reading and sending files without the capability for real-time control.

• The user interface in many of today's controllers is essentially a push-button operator's panel (the new touch-sensitive screens are essentially the same) that does not take advantage of the rich information environment.

• Programming is performed through a cryptic controller-specific G-code language, which is not practical for nontrivial programs and necessitates a special, installation-dependent post-processor to translate from higher-level parts and tool programming languages such as APT.[5]

---

[5] In principle, G-codes are the subject of a standard (EIA RS-274) and are not supposed to vary from controller to controller. But different implementations of the standard for different controllers lead to incompatibilities between the G-codes for them.

The challenge in designing a controller is providing the appropriate computer environment for the integration and implementation of a complex machine tool environment that applies these individual technologies. This challenge cannot be met with current controller technology. Future machine control technology should be characterized by two central themes:

1. *A self-sustaining work environment.* A self-sustaining machine, for any kind of processing operation, will be serviced in its immediate vicinity by a dexterous manipulator or other automatic loading device, dedicated to the continuous needs of the process, such as material supply and unloading. Instructions for the machine (e.g., cutting path) will be generated directly from parts specified by CAD/CAM design. A variety of on-machine sensors will provide vision, touch, force, and temperature senses, with the task of recognizing unexpected events, performing in-cycle inspection, and adjusting the production parameters.

2. *An open system for real-time control and communication.* An open system is one in which the control or communications product developed by a third party also adheres to the standard on which the open system is based; a good paradigm for open machine control and communications architectures is represented by the standards that characterize UNIX and TCP/IP. Machinery controllers for unit processing such as MOSAIC (Box 4.1) aim to promote this atmosphere. If successful, the U.S. machine industry will grow as market opportunities are expanded for sensor companies, diagnostic software developers, and all ancillary product suppliers.

The following research and development areas should be addressed in order to achieve the vision of 21st-century control of factory equipment:

• *An open architecture for machine controllers.* An open architecture would provide a more efficient and "user-friendly" environment for operation and programming, the ability to integrate various devices with a machine's construction and operation, and the ability to communicate more tightly with CAD/CAM systems and factory-wide operational management systems. An open architecture ensures that all components share the same high-level operating system, programming environment, communication facilities, and other computer resources. It accommodates the installation of new devices and sensors as a part of the machine-specific configuration. Controller architectures and software must also support communication with "upstream" functions (including computer-aided process planning and computer-aided design) and "downstream" functions (including those that drive and process sensor data). Also, such an architecture has to be compatible with a range of very different factory characteristics (fixed versus variable product flow, static versus rapidly changing product mix, shared versus assigned tooling resources, and so on).

**BOX 4.1 An Example of an Open Architecture
for Machine Controllers**

The MOSAIC project of the late 1980s (Greenfeld et al., 1989) was one of the first attempts at establishing an open architecture for machine control. MOSAIC provides all of the expected functionality of a conventional "closed" computer numerically controlled (CNC) machine tool. That is, the parts-programmer of a CNC machine tool can interact with the MOSAIC system from any automatic program for machine tools (APT)-like, CAD/CAM terminal with the result that the files generated through this interaction are postprocessed into machine-level commands that provide exactly the same functionality as G&M codes.

However, MOSAIC adds a computer operating system that serves as a uniform platform on which CNC applications can build, specifically, a real-time version of UNIX. The architecture of this operating system allows a set of independent application programs to be developed. This set of application programs can be developed by any third party and brought into the MOSAIC controller for use by the local programs or system developers.

By contrast, a closed architecture constrains local programmers to work with the predefined set of G&M codes that are supplied with the machinery company's vendor-specific controller (Fanuc, Mazak, and Cincinnati-Milacron are today's most-often-seen controllers). Recently, some of these vendors have also claimed an "open" system; but close examination usually shows that the "openness" is only in relation to their own expanded library functions, written in local formats. They will not be "open" to any arbitrary third-party software developers able, say, to supply new routines for new sensors coming onto the market.

In general, machines that plug into an open system must be equipped with a general-purpose computer environment and bus structure that in a "local" sense will control the axes of motion and the various sensor-based devices and will manage programs and data locally. However, to be part of a "broad" intelligent environment, the machinery must use communications and networks that are universally accepted in the computer culture. The machine should be adaptable to the changing environment and tasks, and thus, modular, in terms of its controller's computer configuration, and in terms of the mechanical construction.

MOSAIC is suggestive of what an open architecture controller might be. Nevertheless, any commercially viable product would have to deal with interactions between the large number of controller features required in such products.

• *Advanced manufacturing languages for unit process programming.* While existing languages (such as APT and Compac for machining) must be supported, a more flexible language is needed. It should include provisions not only for real-time control, but also for the operation of accessory devices in conjunction with the machining process, a more direct connection to CAD/CAM systems, and a flexible interface for user applications. Such a language might also provide controllers with information defining the work being performed, rather than simply specifying the motion of a point through space (as is the case for most current languages).

• *Appropriate data and knowledge structures.* Different needs and different local working environments require different data and knowledge structures and different ways of managing these structures. For example:

• Special real-time database systems are required for real-time acquisition and control of data, since conventional database systems do not have the responsiveness necessary to support rapid acquisition of data. For example, data may need to be time-stamped, so that events can be synchronized at a later time in a conventional computing environment. In addition, databases for ostensibly different purposes must often couple to each other: databases related to equipment maintenance are different from databases that support lot-scheduling needs, but since schedulers need to know the status of equipment, the equipment maintenance database must be linked to the scheduler.

• Expert systems today tend to be rigid in the sense that they are not readily interoperable with another knowledge base derived from the expertise of a different expert. Research is needed to find ways to reconcile knowledge bases derived from different experts on the same subject. In addition, knowledge-based technology must be developed that is responsive enough to be used with real-time systems.

• Databases that pass and manage information about events and activities on the shop floor are needed to inform design engineers as well as shop floor workers handling successive shifts. Design engineers may not understand very well what is actually happening on the shop floor; expert system shells have been useful for presenting operators with theoretical best practices and capturing their immediate reactions, which are then routed instantly to engineers along with other information that explicitly defines the context of an operator's remark. For shop floor workers, special databases are needed that will track on the fly newly discovered and ad hoc information (e.g., flaws discovered on the midnight shift) for use in the next shift.

• Databases and knowledge representations that enable decision makers to distinguish between unusual system behavior that nevertheless reflects "reasonable" or "appropriate" behavior and unusual system behavior that reflects some kind of failure. As factory managers become privy to larger amounts of information, such questions will arise more frequently.

• *Validation of models underlying controller designs, for example, a controller based on an expert system that interprets sensor input.* The relationship between different inputs and different rules in an expert system needs to be well understood if operators are to have confidence in the operation of the controller. Thus, reliable techniques are needed to demonstrate the validity of the underlying expert system model.

• *The human-computer interface (HCI)*. Given the complexity of a factory's information environment, an effective factory HCI must provide displays that will provide a human being with the appropriate level of detail for his or her needs (which may change as more is learned about a given situation). Desirable features for a factory HCI include:

> • A direct, easy-to-use interface into product design databases so that product specifications can be easily reviewed on the plant floor;
> • Support for unusual input/output options necessitated by the factory environment. For example, an operator with both hands full may need to interact by voice, or an information display may need to be viewed at a distance, or a very concise set of keystrokes may be needed to save time in user input; and
> • Tools that enable the end user to tailor the interface for individual needs and to standardize the interface across multiple machines. The varying HCIs of differing equipment currently present significant obstacles to training employees and attaining quality in manufacturing.

• *Design of shop floor equipment for repairability and the ability to work around equipment "crashes," including diagnostic software*. Factories are fragile environments; small changes can cause large disruptions in work flow. Such disruptions can be very expensive; 1 hour of shutdown can cost upwards of several hundred thousand dollars. Thus, problems and disruptions must be diagnosed quickly, and fixes implemented promptly. Quick fixes may be applied to bring crashed equipment up, even as more permanent remedies are studied and adopted.   Many capabilities related to prompt diagnosis and repair will be achieved by using sophisticated software, databases, expert systems, and networks. The requirement will be speed and accuracy; a quick but poor solution may be more costly than a slower but more correct one. Study of the trade-offs between the time for and the completeness of solutions is a worthy area for manufacturing research, as is the designing of work flows in factories so that there are no single points of failure. (Today, such failure points include network servers, central databases, and cell controllers.)

• *The nature of real-time requirements*. Factories operate in real time, and a controller that issues a command too soon or too late may cause inadvertent damage. But real-time computing and control are extraordinarily complex and demanding. Thus, several areas require attention:

> • *The role of time in various control situations*. Delay times owing to computer operation cannot be predicted on the basis of simple formulas as they are in the case of analog circuits; rather, they are currently a matter of best guess and experiment. Even then, they are subject to considerable statistical variation. Research in this area should seek to

establish the characteristics of the time distribution that will lead to successful control, characterized by the distribution of response times, especially the mean and average times and the worst-time behavior.

• *Mathematical formalisms for real-time control.* Developing appropriate formalisms for representing and controlling devices is key to the development of good real-time systems. Today, classical control theory is the dominant formalism used in real-time control, although modern control theory, state space analysis, fuzzy logic, and neural networks have made inroads to some degree. Controllers of the future are likely to involve all the control techniques—classical, modern, fuzzy logic, and neural networks—in an integrated approach to control.

• *The development of a real-time operating system for manufacturing,* suitable for the very high speed control required for unit processing operations. Current general-purpose operating systems do not provide the response time essential for maintaining the speed, accuracy, and safety features needed in future machinery. A real-time operating system for manufacturing should be compatible with industry-standard operating systems such as UNIX or MS-DOS with respect to high-level management, file-system operations, communications, and programming environments. Capabilities for network-based coordination with other tools running on the same real-time operating system will also be necessary for the integration of equipment controlled by separate controllers.

• *Wireless communications.* Since computing will be used throughout the manufacturing environment, and it is impossible to predict beforehand where these interactions will occur, wireless communications will probably be required. Wireless communications will make factory reconfiguration easier, faster, and cheaper—as the locations of equipment need not be tied to communications portals—and will be useful in supervision, maintenance, handling of materials, and other activities in which personnel cannot be tied to fixed workstations. But the trade-offs associated with wireless communications in a factory environment are complex, for example:

• Trade-offs between the capital costs of fixed building wiring with very high bandwidth versus the facility costs of wireless transmitters and receivers that operate within a very restrictive bandwidth;

• Trade-offs among broadcast power, geographical coverage, and multi-station interference; and

• The potential hostility of the factory environment to such communications (e.g., wireless communications in the presence of radiation-sensitive equipment).

Research will be needed to support a rational decision-making process with respect to these trade-offs. Research will also be needed in the area of the transmission and reception of complex and varied signals in a sensitive environment. Other relevant research (and standardization) would address such issues as maximizing the use of the limited electromagnetic spectrum and spread-spectrum-type wireless communications using trellis-type encodings to ensure maximum security, reliability, and noninterference with sensitive manufacturing equipment.

Beyond research, standards (for components, configurations, interconnections, and control software) and related assistance (e.g., from Sematech activity for semiconductor manufacturing) are needed to ensure consistency and interoperability of implementations, to coordinate and standardize open systems concepts in control systems, and to motivate system and equipment suppliers to use these mechanisms in their equipment.

Additional dimensions of equipment controllers are discussed later in this chapter under "Facilitating Continuous Improvement" for reasons that are made clear in that section.

Sensors

To ensure the optimum performance of production processes, it is necessary to have information on what those processes are actually doing in real time and how those processes are affecting equipment, tooling, work zones, material handling, and workpiece material. Knowledge databases and predictive models of a process and its components provide a baseline of "in-advance" information, but if routine deviations are greater than can be tolerated for production within tolerances, sensors are needed to augment this baseline information for process monitoring and feedback to controllers.

In existing systems, estimates of the impact of sensing systems on process performance indicate as much as a sixfold increase in effective operation speed (Eversheim, 1991). Add to that the increased yield due to prevention of defective workpieces and the impact is more impressive. Moreover, sensors can also support worker safety, prevention of damage to a machine, prevention of rejected workpieces, prevention of idle time on the machine, better interfacing of transfer mechanisms with equipment, and optimal use of resources.

The increasing demands of unit processes have encouraged the development of systems using a variety of sensors. Examples of devices and processes into which sensors will be incorporated include extrusion dies for control of temperature and metal flow and surface finish; turning tools, for thermal control to provide maximum life and load; and continuous processes, for transmitting process parameters. Sensors will interface so closely with processes that complicated external interpretation of data will no longer be necessary. Future sensors will provide diagnostics (as many do today) as well as fault tolerance and in some cases self-healing capability. Sensors will provide salable sample rates for analysis of process variation, summary-level information, and statistical descriptions of process parameters.

Sensors can play an important role in many aspects of production, including:

* *Quality control.* Many quality control schemes rely on pre- and post-process inspection—that is, catching a mistake before a component enters an individual unit process (when nothing is happening) or when it exits from the subprocess (when the unit process is completed). Extension of quality control to in-process inspection, so that problems can be caught in "real time" and perhaps corrected before much additional processing, is even more dependent on sensor performance and the availability of advanced sensors.
* *Workpiece placement.* Machining tools depend on information about tolerances, orientation, material characteristics, and assembly sequence. Sensors are needed to assess these characteristics before the partially finished assembly or part is worked on.
* *Use of intelligent processing equipment.* A unit process could do more or less to a part being fabricated (e.g., heat it more, remove less metal) depending on how the particular part being worked on responded to its treatment. A piece of equipment with location sensors could signal its location as it was moved around on the shop floor.
* *High-precision fabrication.* High-precision fabrication is characterized by very stringent tolerances on form, dimension, or surface features in the presence of a number of material, tooling, and environmental variables. As a production tool shapes material for a finished (or intermediate) product, the reaction of the material to the tool has an influence on what the tool should do next. For example, a machine that unexpectedly encounters more resistance than anticipated to a given cut may need subsequently to exert more force, and a sensor is needed to close the feedback loop to the tool's controller so that the necessary orders may be given.
* *Use of tools.* A sensor-loaded tool moving along with the process could provide information on product and process specification compliance and location. The new active sensors will change the way products are transported, tracked, monitored, and maintained at every stage of their existence—from creation to recycling to disposal. Researchers have developed miniature vibration sensors and accelerometers that, if appropriately built into a machine structure (like the resin concrete structures now in use as machine tool structures), could provide in situ sensing capability for control of machine stability and deflection.

Sensors are key elements in the other enabling technologies of process control and process precision and metrology. A vast array of sensors are commercially available or under development in research laboratories; these sensors support unit processes through an extensive array of sensor technologies ranging from optical, infrared techniques to high-frequency ultrasonic and acoustic emission.[6] Sensor innovation will be necessary in the future, because many characteristics of machining processes today (e.g., very low power consumption) tend to reduce the effectiveness of, or render useless altogether, a large number of traditional sensing methodologies based on force, torque, motor current, or power measurement. Examples of more novel sensors include acoustic sensors to monitor and improve tool performance and surface finish or to monitor the relative location of parts in a finished product; photoelectronic and ultrasonic sensors that enable intelligent processing equipment to position and set functions; and bar code and radio frequency sensors that identify and report on the status and location of parts. In many cases, it is likely that relevant sensor technology will have been first developed for application in other fields.

The first major contribution of information technologies to sensors was the idea of digitized output, which removed analog variation from the outputs. Later, self-calibration and error detection were added to allow self-diagnostics to assure that the information from the sensor was not detrimental to the process control. Sensors for the future production environment pose a number of challenging information technology research questions:

• *Standardization.* Today's sensors (and actuators) require customized drivers for their operation, greatly increasing the cost of an otherwise inexpensive sensor or actuator. Architectures and interface standards are needed so that users can select from a catalog of sensors and actuators, each with a certain limited set of parameters specified in the catalog, and plug those selected into a common control system with only minor, automatic configuration. When such architectures and standards are in place, sensors should be able to feed data directly into a process control system. Sensors connected through such architectures would be linked directly into databases for dynamic updates usable by machine controllers.

• *Sensor characterization.* An appropriate characterization would describe a sensor or actuator at various levels of abstraction (e.g., as an abstract physical process, as an electrical device producing an output, as a "smart" device with a certain amount of preprocessing capability, including accuracy corrections, zero-offset correction, and perhaps self-calibrating and self-scaling functions). The method of characterization should be applicable to a wide range of devices, including sensors from simple thermocouples up to image systems, and actuators from simple on-off switches to multiaxis robots.

• *Sensor fusion.* Many sensing techniques are not individually reliable

---

[6] An excellent review of these sensors as related to machining is provided in Shiraishi (1988; 1989a,b).

enough for process monitoring over the normal range of process operation. Multiple sensors that are collectively effective over the entire operating range may be enablers for effective real-time process monitoring if questions of integration can be resolved. To be effective, a multisensor approach requires more attention to feature extraction, information integration, and decision making in real time.

• *Sensors with on-board intelligence.* With on-board intelligence, the collection of data might be collapsed into a statistical distribution (for routine data) and specific reports of exceptions to the distribution, thereby reducing bandwidth requirements for reporting sensor data. On-board intelligence might enable a sensor to recognize physical features of a machined part (e.g., color, size, shape) to detect the presence or absence of a component or characteristic of a part. Intelligent sensors also might include capabilities for signal conditioning and processing (reducing the impact of uncertainty in sensor data), and perhaps models relating measured values to monitoring and control variables and strategies for using the information gathered.

FACILITATING CONTINUOUS IMPROVEMENT

When applied to the shop floor, continuous improvement refers to the continual monitoring of shop floor practice and how that practice is reflected in the products that result. Whereas traditional shop floor environments emphasize the desirability of procedures that are consistent and hence unchanging, continuous improvement suggests instead the incremental evolution of procedures to make products of ever-higher quality and timeliness. Products are monitored for timeliness and for quality at every stage in production; the resulting information is passed back to process operators with enough detail to suggest corrective actions that may need to be taken to improve the process on the next round. Process operators implement certain changes and then see if their changes have had the desired effect. If so, such changes are recorded as annotations to the then-current operating procedures to effect the improvement. In this manner, continuous improvement relies on standardization of practice, resulting in an empirically relevant point of comparison against which to measure change in practice, rather than standardization (and hence invariance) of procedure.

At the heart of continuous shop floor improvement is the idea of operator-centered control systems that enable operators to take a wide range of actions that may be necessary to improve production quality or speed. Research is needed on several dimensions of operator-centered control systems:

• *Capture of most recent agreements.* Annotations to standard operating procedures can be described as the most recent agreements (i.e., details of understanding reached on the last shift) between operators and the most recent version of those operating procedures. Such agreements must be passed on to subsequent shifts if the knowledge is not to be lost. When captured, these agreements should specify the nature of the changes in operating procedures indicated, the rationale for these changes, the changes in output to be expected, and the circumstances under which these changes are to be implemented. A particularly promising approach is the use of expert system shells that allow informed agreements to be captured on the spot using the operator's expertise.

• *Effective diagnostic capability.* A tool should be able to report its state of health: where the process it implements is weak, what problems are impending, and when action will need to be taken.

• *Communication with other shop floor systems.* This connection would provide, for example, scheduling that reflects the most recent priorities so that the most important jobs could be done first. A connection to the material-handling system that feeds the tool would inform the operator of unexpected delays in the arrival of materials and permit, for example, maintenance to be performed in that unexpected free time. Information about the tool's output would be available to workstations downstream.

• *User-configurable interface.* A display (control panel) should be dynamically modifiable in accordance with user preferences and needs. Once a change called for by a new agreement has been implemented, the user should be able to modify the interface to reflect that change.

• *Administrative control.* Operator-centered tools should account automatically for operator time and attendance, labor distribution, production counts and rates, quality levels, and similar administrative data. Greater accuracy and precision in obtaining these statistics should be possible.

All of these elements are aspects of the equipment controller. But they are discussed separately from the earlier discussion of equipment controllers in this chapter to underscore their importance in capturing information generated at the back end of the production cycle.

CONTROLLING AND MANAGING PRODUCT CONFIGURATION

Manufacturers often produce variants of a given product. For example, Intel Corporation produces microprocessor chips that are "upwardly compatible" in variants known as the 386, 486, and Pentium series. Each model also has variants (e.g., those with math coprocessors versus those without, those with power-management features versus those without, and so on). Production operations that can shift easily and rapidly from producing one variant to another would give manufacturers a needed degree of flexibility in their management.

Flexible production of this sort is likely to depend on the implementation of two ideas of different time scales. The first is that of a single production facility whose internal operations can be modified substantially by software-based changes to machine controllers and schedulers.[7] The second notion is that of a manufacturing operation that can draw globally dispersed units providing specialized expertise and unite them temporarily to carry out a specific production task (e.g., to produce a specific product or product line) and then disband them when the task has been completed.

To effectively control and manage product configuration, research is needed to manage product variations and to check the validity of proposed configurations. At present, companies develop product-specific systems or may use artificial intelligence techniques (also often product-specific themselves) to deal with configuration control. In addition, configuration control and management require a universal product configuration language and methodology that can be used by both marketing and manufacturing personnel to develop valid product configurations.

SPECIFIC RESEARCH QUESTIONS

As in Chapter 3, a variety of research questions are motivated by discussions earlier in this chapter. But the caveat posed in Chapter 1 also applies here: research to fill the gaps in the scientific and engineering knowledge about shop floor processes to be supported by information technology is essential if the promise if IT is to be exploited fully. For example, a deeper basic understanding about the physics of part-tool interactions or the physics and chemistry that might underlie possible faults in a process will be necessary if IT is to control tools or to help diagnose faults in a production line. As always, IT can help immeasurably in exploiting knowledge and information that people have, but by itself it is not a substitute for that knowledge and information.

- How extensive should any factory system be? One hierarchical system

---

[7] Even an idea as simple as changing the order in which variations of a product are manufactured can have significant effects on productivity. For example, an AT&T manufacturing plant was responsible for the production of 80 PBX (private telephone exchange) systems each week in a wide range of sizes with many different options. The order of manufacture was determined by the sequence in which orders for these systems were received; this pattern resulted in a large and undesirable variation in flow rates of work through the plant as the result of the many different feature sets needed. By rearranging and thereby improving the manufacturing sequence of these 80 systems, a significant improvement in throughput times from the lines feeding final assembly was achieved with no additional equipment purchases. See Luss et al. (1990), pp. 99-109.

with many embedded applications for specific purposes, or a set of distributed applications with few interactions?

• How complex may the user interaction systems be before the complexity overwhelms the user?

• How are new systems introduced into a factory that contains many older or even obsolete legacy systems? How can an existing, obsolete system be replaced with minimum factory disruption?

• How are decision-making systems transferred into decision-making roles?

• What level(s) of abstraction are appropriate for controlling factories?

• What is the trade-off between levels of detail required and the speed of response?

• What characteristics and configurations of autonomous agents give rise to emergent behaviors, swarm stability, and divergent behavior?

• How will humans interact with factory systems? What will be the nature of the user interfaces?

5

Modeling and Simulation for the Virtual Factory

INTRODUCTION

The development of manufacturing systems often proceeds in the operating factory by trial-and-error experimentation using valuable factory resources and time, with a resulting high cost in lost manufacturing productivity due to prototyping and debugging on the factory floor. Data continue to be generated at an exponentially increasing rate, but there is little opportunity to assimilate, much less act on, the information they represent. The abundance and variety of customized and unintegrated new technological capabilities introduce problems for managers, engineers, supervisors, and operators within a factory, with which the work force may be ill-prepared to cope. Better ways to introduce new manufacturing technology onto the factory floor must be developed to help improve our manufacturing capabilities.

As noted in Chapter 4, the factory environment is complex. To develop high-level insight into what is happening in the factory environment and how factory operations might be affected by changes in that environment, manufacturing researchers have attempted to construct simulations and models of factories that factory decision makers can experiment on without disrupting actual production. This approach requires a virtual factory to be a simulation that faithfully reflects that operation in all of the dimensions relevant to human managers, from the week-long and month-long time scales that characterize today's material requirements planning (MRP) and manufacturing resources planning (MRP-II) planning and scheduling systems to the second- and minute-long time scales that characterize real-time dynamic schedulers (human or computer) of shop floor operations.

CONSIDERATIONS IN THE DEVELOPMENT
OF VIRTUAL FACTORIES

Learning from Past Problems

The idea of a virtual factory is not new. But in the past, the development and application of large coherent models of factory operations have been largely unsuccessful, even as more local simulations (e.g., those operating at the equipment level or process level) have succeeded. The models of the past have not had the content to capture the essence of factory operations, yet even these inadequate models were so difficult to produce that after great effort they typically ground down into "analysis paralysis" and were abandoned.

Possible causes for such failure include the following:

• The initial models were too poor in detail to provide answers that satisfied factory demands, and so the concepts were abandoned. Even small events may produce large fluctuations in factory operations, and adequate models must be capable of reflecting these subtle influences.

• The simulations were too slow to provide timely answers. Timeliness is paramount in factory operations; a piece of equipment that is unexpectedly down, or is being used for another task when required to perform a function, can disrupt a shift schedule. Data must be provided and acted on very quickly to preserve the integrity of operations.

• The representations of the process were inaccurate, leading to wrong answers. Processes may be simple or complex; in any event, all the important characteristics of a process need to be accurately represented so that applying a model will supply useful answers.

• The user interfaces were so complicated and/or incomprehensible that they were unusable. The most common user of information is a human being, who can be overwhelmed by either the number or the complexity of the user interfaces required to access or disseminate information. For example, senior factory managers are best able to comprehend results that are explicitly tied to financial metrics of performance; results tied to metrics relevant at lower levels in the hierarchy will be less helpful to them.

• There were insufficient skilled personnel to understand and apply the models intelligently. Use of models is not inherently easy. It requires skills that enable using models and simulations, as well as understanding and analyzing the results. Learning these skills requires education and training.

• There were sufficient skilled experts on the factory floor to manage operations, so that modeling and simulation were considered unnecessary. When things are going relatively smoothly, new tools such as modeling and simulation are felt to be superfluous; current skill sets are thought to be sufficient to do the job.

- Invalid input data to simulations led to incorrect results. For example, communications on the factory floor were insufficient to keep the model up to date. (Factory floor communications are used, among other things, to keep track of equipment placement on the shop floor, a factor with a big impact on work flow; if a model does not reflect changes in equipment placement (which may change on a time scale of hours), the model's output relevant to making work flow decisions may well be incorrect.)
- Factories themselves constantly change (e.g., new or modified manufacturing processes may be installed), and a lack of synchronization between a model and what is actually being done on the shop floor may invalidate the model.
- Simulations may have been performed at inappropriate levels of detail for addressing problems of interest and relevance to decision makers. An element of considerable importance in constructing a simulation is understanding what kinds of questions need to be answered for what purposes.
- Today's simulation models are difficult to expand, and they present very difficult problems in scaling up.

For such reasons, factory-level modeling has never succeeded in capturing the attention of senior manufacturing management the way that process and product models have. However, recent advances in information technology make the idea of realistic simulations of factory operations much more feasible than they have been in the past.

Determining the Requirements for Effective Factory Models

A virtual factory model will involve a comprehensive model or structure for integrating a set of heterogeneous and hierarchical submodels at various levels of abstraction. Each submodel will be designed for a specific purpose, but together they will operate from a common source of data or knowledge base and will be able to deal with the task at hand without expensive or time-consuming hand-tailoring of interfaces for a user's particular needs. To a very high degree, the software used to control actual factory operations will also drive the operation of the virtual factory model, although it will do so very rapidly so that simulations can be run on a timely basis.

A key consideration in the development of these tools will be accounting for the essential stochastic nature of events on the factory floor.[1] Since managers

---

[1] *A Research Agenda for CIM* (MSB, 1988, p. 20) recommends research on methods for representing decision objects, for example, states, actions, utilities, prior and posterior probabilities, samples, costs, and decisions; methods for mapping these to manufacturing objects; development of relevant heuristics and algorithms; and exploratory assessment of the merit of these techniques in specific domains.

will be using virtual factory models to perform "what if" exercises to anticipate unexpected events (e.g., tool breakdown, new demands for expediting orders, the arrival of new machines, unplanned maintenance) and to indicate what actions need to be taken (e.g., which machine is to be changed, what materials are to be provided), a simulation based only on deterministic factors would correspond to one "experiment," and in a stochastic environment, there is no reason to expect that the same line configuration or schedule plan would yield the same result on any particular run. However, a suite of modeling tools that could test a given configuration or plan against thousands of experiments (and the random events in each run) might well provide a meaningful basis on which to make decisions about a configuration or plan. A second important consideration is that the development of simulation models and tools should be broad enough to encompass both the factory technologies and processes of today (which are embedded in the manufacturing infrastructure and may not be replaced for many years to come) and those of tomorrow (i.e., those technologies and processes that have not yet been deployed widely or even invented yet).

A complete simulation of even a modest factory will require the integration of numerous models and is out of reach today. However, an appropriate first step in developing a full factory model is the virtual production line, which involves simulation of individual tools and, more importantly, simulation of the integrated operation of the tools in production. A wide variety of individual, single-activity models are already in use in manufacturing. But the comprehensive integrated modeling of the manufacturing enterprise will provide new insight into the causes of and remedies for scheduling bottlenecks and new strategic options.[2]

An ultimate goal of the research agenda outlined here would be the creation of a demonstration platform that would compare the results of real factory operations with the results of simulated factory operations using information technology applications such as those discussed in this report. This demonstration platform would use a computer-based model of an existing factory and would compare its performance with that of a similarly equipped factory running the same product line, but using, for example, a new layout of equipment, a better scheduling system, a paperless product and process description, or fewer or more human operators. The entire factory would have to be represented in sufficient detail so that any model user, from factory manager to equipment operator, would be able to extract useful results.

There are two broad areas of need: (1) hardware and software technology to handle sophisticated graphics and data-oriented models in a useful and timely

---

[2] *A Research Agenda for CIM* (MSB, 1988, pp. 13-15) recommends research on methods and technologies for process operation analysis, optimization, control, and quality assurance and (on p. 17) recommends research on methods for coupling process operation with process planning and resource allocation so that these activities can be accomplished concurrently.

manner, and (2) representation of manufacturing expertise in models in such a way that the results of model operation satisfy manufacturing experts' needs for accurate responses.

MODELING TECHNOLOGY

A requirement for manufacturing models is that all levels of detail be internally consistent, since very small influences at a very local level may have a significant impact even at the highest factory level (e.g., if a piece of equipment goes down, the resulting schedule disruptions can shut down the entire factory). A factory may well demonstrate mathematically chaotic behavior, such that very small influences may have totally unpredictable outcomes. The enormous range of objects (e.g., pieces of equipment, people, raw materials, batches or lots of material, process steps, process control, assemblies of objects, a wide variety of information "objects," material-handling and robotic equipment, and so on) encompassed by even a partial manufacturing model further adds to its complexity.

In general, the complexity needed in a model to generate accurate results strongly influences the time it takes to simulate some factory condition. Thus, there is a trade-off between speed of response to provide timely answers and the level of detail required to provide "good enough" answers. For example, some simulations of semiconductor processing operations may take hours or days of computation (i.e., may take much longer than the operations take in reality); such performance may be useful for analytical purposes but not for controlling or advising for real-time operations.

If advanced modeling technology is to be useful in the future, it must not disregard the large investment that has been made in existing data systems. The client-server systems and relational databases of today, and even object-oriented databases, will be the legacy systems three and four decades hence. New technology must be able to fold these legacy systems into its representation and be able to resolve semantic differences that may exist.

In addition, new modeling technology will likely reside in an extensive distributed computing environment. Thus, fundamental issues of maintaining data integrity in a distributed computing environment need to be resolved. Administration of systems, networks, and applications is today resource-intensive; more efficient and effective methods of managing these elements will be required.

Finally, information at the appropriate level of detail must be presented to relevant decision makers and analysts. Visibility of the activities of an enterprise is a prerequisite to effective control. Research is needed regarding how to present to different users (from managers to operators) the information inherent in the business model, which can reflect an actual state, potential future states, and post-mortem analysis of past states to determine what went right or wrong. People should be able to prepare presentations of various kinds of information on their own, rapidly, and without the use of support personnel, and to retrieve "lessons learned" and present that information at the appropriate times to users.

Specific research issues related to developing manufacturing models include improving methods and tools for:

• *Efficiently managing large amounts of related data distributed over many machines and locations.* Data may be directly linked to the computer model of the production process, automatically updating the status of the model, and it will be desirable to automate data collection to the maximum degree.

• *Specifying complex data relationships.* Since the various scales of manufacturing are intimately interlinked, models will require large amounts of detail, probably stored in distributed object-oriented databases. However, these databases will need to be consistent across geographical sites and will span many time zones. Large servers will be needed to manage the model applications and the propagation of changes.

• *Easing interoperability of design process tools.* Interoperability problems lie in two domains, in the meaning of the information passed between tools and in the mechanics of their transfer. To address the first issue, a complete data model of the simulation process must be created, one that will specify the information at every stage of operation. When this is coupled with a flow model of the simulation, it will provide an unambiguous definition of what information is being interchanged and when. The mechanics of information transfer require either the development of standards for passing messages between tools (e.g., the communication required between objects in an object-oriented paradigm[3]) or the specification of programming interface standards, or both.

• *Constructing formal models that reflect factory resources.* Reflecting factors such as resource availability, condition or repair status, and shop floor location, such models should interact with process and product models; indeed, at sufficiently high levels of abstraction, process model components may appear as components of a resource model. Careful configuration management of these models will provide an approach to alleviating the problem of validating models for a continually changing factory environment.

• *Presenting data (e.g., using human interface tools).* For example, model displays may have to be very large, either to enable viewing large quantities

---

[3] This is one stated goal of the product data exchange using STEP (standard for the exchange of product model data) project described in Chapter 3.

of data or to show lots of visual detail. The displays may even have to be mobile, so that they can be carried around in a factory. Display panels may be attached to a computer or be in the form of glasses used by a viewer to simultaneously view the model and the real world.

• *Reducing the magnitude of the testing and validation task problems by carefully configuring the relationships between models used for control and those used for simulation.* A factory control system that has been success-fully tested for operation in real time could in principle be connected to a set of inputs that simulate the input it receives in real time, and could thus serve as the foundation of a (partially validated) simulation model as well.

• *Developing new paradigms for understanding dynamic interactions among the different aspects and components of the manufacturing enter-prise.* A real factory consists of many interacting elements; thus, it is reason-able to expect a high-fidelity simulation to consist of many interacting ele-ments (models) as well. But the real factory is difficult to understand, in part because of these interactions, and so a real problem of understanding is posed by a complex multitude of interacting models. Developing new ways to understand the complexity, nature, and scope of interacting components will be a major challenge with ramifications for ensuring model fidelity and validation.

The following are research areas within modeling technology that need more work:

• *User interfaces.* Since factory personnel in the 21st century will be interacting with many applications, it will not suffice for each application to have its own set of interfaces, no matter how good any individual one is. Much thought has to be given both to the nature of the specific interfaces and to the integration of interfaces in a system designed so as not to confuse the user. Interface tools that allow the user to filter and abstract large volumes of data will be particularly important.

• *Model concurrency.* Models will be used to perform a variety of geo-graphically dispersed functions over a short time scale during which the model information must be globally accurate. In addition, pieces of the model may themselves be widely distributed. As a result, a method must be devised to ensure model consistency and concurrency, perhaps for extended time periods. The accuracy of models used concurrently for different pur-poses is a key determinant of the benefits of using such models.

• *Testing and validation of model concepts.* Because a major use of models is to make predictions about matters that are not intuitively obvious to decision makers, testing and validation of models and their use are very difficult. For factory operations and design alike, there are many potential "right answers" to important questions, and none of these is "provably correct" (e.g., what is the "right" schedule?). As a result, models have to be validated by being tested against understandable conditions, and in many cases, common sense must be used to judge if a model is "correct." Because models must be tested under stochastic factory conditions, which are hard to duplicate or emulate, outside the factory environment, an important area for research involves developing tools for use in both testing and validating model operation and behavior. Tools for automating sensitivity analysis in the testing of simulation models would help to overcome model validation problems inherent in a stochastic environment.

• *Model evolution.* Models are not static, in the sense that as users learn more or as the factory environment changes (e.g., as a result of new rules, new heuristics, changing equipment sets, and so on), models have to be updated. In addition, as new knowledge is developed, models need to be enhanced, speeded up, and made more detailed. The information that can be obtained from a model depends fundamentally on the model's capabilities, which are expected to improve with time. This is a fruitful area for research.

• *Chaos theory.* To the extent that manufacturing is a mathematically chaotic environment, chaos theory may be able to show where critical assumptions break down and where they may lead to computationally impossible situations and totally unpredictable behavior.

REPRESENTING AND CAPTURING MANUFACTURING EXPERTISE

It is obvious that the modeling sophistication implied by the preceding section will demand an extraordinary degree of manufacturing expertise at all scales of operation (from shop floor operator to senior management). But a number of factors specific to the manufacturing enterprise exacerbate these demands even further:

• Since activities and actions at the different scales of manufacturing operation cannot be considered independently, expertise relevant to one level will often couple strongly to expertise at a different level. Since models will pass information up and down the management ladder, information exchange mechanisms must be consistent. Hence, a consistent model representation language is needed that will precisely define a model's objects and the relationships between objects at different levels of abstraction.

- The desired coupling of a model (the virtual factory) directly to a real factory demands that changes in the real factory be quickly incorporated into the model, or, conversely, that the model be used to drive changes in the real factory. For example, the movement of a piece of equipment in a model would cause a robot to relocate the corresponding real piece of equipment. Or, a change in the temperature in a model object would cause the real temperature to change in the real factory. To enable such coupling, the links to the real factory from the model need to be very robust and bidirectional.

- Different "windows" into a model will have to be provided for different users, including "help"-line operators, engineers, managers, supervisors, technicians, schedulers, human resources personnel, and maintenance personnel. Each type of user will have his or her own perspective on reasons for using the model, and the model should be able to be tailored to accommodate those specific needs.

It is likely that traditional approaches in artificial intelligence with respect to knowledge engineering will be helpful in capturing expertise, since heuristic or even analytical manufacturing information is difficult to formalize. At the same time, research in the area of capturing manufacturing expertise should not be limited solely to the artificial intelligence knowledge-engineering approach. For example, it would also be valuable to investigate how people involved in the manufacturing enterprise construct meaning when they are listening to others describe some manufacturing process or activity. Hard experience in such situations suggests that even when all participants share the same spoken and written language (English plus mathematics), it is often difficult to specify certain procedures without ambiguity.

The following describes some of the research and development needed for capturing and representing information:

- *Gathering expertise.* Although it has been possible to gather knowledge from experts in certain narrow and well-focused domains, acquisition of knowledge is still difficult. Since manufacturing is such a varied and complex activity, and since the information exists over a very wide range (from high-level factory decision making to running a specific piece of equipment), capturing the tacit knowledge of an experienced work force for later reuse represents a key challenge.

- *Languages for gathering and representing knowledge.* An information-theoretic based language is needed in which to gather and assemble responses from experts and channel them into a machine-usable form. In addition, a good language is needed to represent the information in the model. This may or may not be the same language as that used to gather knowledge; however, it will certainly be closely related.

RESEARCH AREAS NOT SPECIFIC TO MANUFACTURING

A number of research problems in areas not specific to manufacturing will also need to be addressed if the virtual factory is to be realized. For example:

• Models that must be concurrently activated and used may have to be built by geographically dispersed teams. Methods to provide for concurrent engineering and for keeping models or pieces of models synchronized will have to be developed.

• Models must run at real-time speeds if they are to be used in conjunction with real-time factory operations; they must run 1,000 or 10,000 times faster than real time if they are to be used to experiment with different factory configurations.

• Testing and validation of models will be difficult, since there are so many variables and combinations of objects involved. "Provably correct" modules will have to be developed, so that large models can be constructed that are free from error. Tools to help build models will be required.

• Model security, to protect a model from unwanted tampering or accidental incidents, to make the model accessible to users but protected from unwanted actions, is another area of concern.

• The computational requirements for such simulations will be very high, perhaps requiring the use of supercomputers. However, traditional supercomputers have been configured for homogeneously parallel problems. The virtual factory simulation will require extensive parallel computation, but it will not be homogeneous. Therefore research is needed on new techniques to apply massive parallel processors to the simulation question.

6

Information Infrastructure Issues

INTRODUCTION

Manufacturing information infrastructure refers to the computing and communications facilities and services needed to facilitate, manage, and enable efficient manufacturing; key elements include database and information management systems, data communications networks and associated services, and management of applications software. For the next generation of manufacturing, information infrastructure must have the high degree of connectivity, compatibility, and ease of use that already characterizes traditional physical infrastructure (electric power, water and sewage services, telephone service, and the like).

A vision for the manufacturing enterprise of the 21st century includes the ability to use an integrated information system to support the entire product cycle from product design to product delivery to the end user. Such a system would facilitate communications among groups, automate "corporate memory," and aid in decision making. Likely using the National Information Infrastructure (NII), this system would function at both the intrafirm (e.g., connecting branch offices dispersed all over the country) and interfirm (e.g., connecting the firm to suppliers and customers) levels. Such opportunities are particularly relevant to small manufacturers that will be able to find niches for their specialized capabilities more easily when the NII enables convenient electronic connections for firms small and large.

The use of networked infrastructures can diminish the role played by place, time, and hierarchy in the structure and management of organizations. A wide range of organizational possibilities remain to be fully explored, and thus inves-

TABLE 6.1 Research to Advance Infrastructure Systems

Subject Area	Example of Research Needed
Architectures and standards	Standard manufacturing control architectures Generic functionality within control architectures Cost-benefit criteria for potential standards
Intra-enterprise and inter-enterprise integration	Principles and architectures for coupling network-based applications Automatic interpretation of transactions Automatic message routing and associated processing Support for multiple protocols and multiple speeds over a given medium Time-critical message delivery in interconnected networks Protocols and services that support specific demands of object-oriented applications Services for human- and machine-based browsing and searching of information and resources Tools and techniques supporting supply-chain dynamics and associated planning Mechanisms and systems to support information session management
Architectures for autonomy and distributed intelligence	Autonomous agents to monitor and respond to production events Knowledge agents for enterprise-wide management of models, names, transactions, and rules Architectures for manufacturing systems involving distributed intelligence Tools to find and distribute information Stable sets of rules for interacting agents Dynamic variations in agent autonomy
Information retrieval systems	Modeling and prototyping functions for user interfaces Next-generation data manipulation languages Maintenance of data consistency and integrity through database updates Network-based services for information browsing and searching
Software engineering	Simplification of system designs, operation, and maintenance Self-healing systems Support for new programming paradigms Tools for component-based architecture life-cycle approaches General tools unconstrained by the limitations of specific programming languages Tools with aspects of knowledge-based collaboration software

TABLE 6.1 continued

Subject Area	Example of Research Needed
Software engineering (*continued*)	Rapid prototyping and other methodologies for faster development
	Techniques for encapsulating legacy systems and developing mediator support
	Analysis methodologies, metrics, and selection techniques
	Reference architectures that cut across manufacturing domains to support software reusability
	Data representations that depict both spatial and temporal aspects
	Tools for better and more user-configurable user interfaces
Dependable computing systems	Better technology to support changing software without removing a system from operation
	Continuous availability of on-line services
	Fault-tolerant hardware and software
	Increased system security and trustworthiness
Collaboration technology	Software, user interfaces, and hardware to support cooperative work

tigation into new information technology (IT)-enabled organizational forms is an area of research interest, as new organizational models will ultimately be needed, or earlier paradigms for understanding organizations revised, to better understand how activities might be reorganized to exploit information technology for better (or higher) performance.

Progress toward this goal involves development of both physical (e.g., wires, computers, gateways, and switches) and "soft" (e.g., protocols and information services) information infrastructure technologies. Areas with research opportunities are listed in Table 6.1 and are discussed in the remainder of this chapter. For convenience, these topics are divided into areas specific to manufacturing (architectures and standards, and integration of activities within and among enterprises) and not specific to manufacturing, although the dividing line between manufacturing-specific and nonmanufacturing-specific categories is particularly fuzzy for infrastructure elements.

ARCHITECTURES AND STANDARDS

Architectures and standards provide structure for many levels of interconnection, from those among devices to those among businesses. There are architectures for systems supporting specific functions, for enterprise-wide information flow, for control of processes, and for management. As the need for exchanging data and knowledge among various manufacturing applications grows and the sophistication of individual applications increases, the interfaces between them become fuzzier and more complex and important. Although each interface may itself be easy to use, correct, and robust, the complex of interfaces in a large, diversified factory is fragile and is subject to failure for relatively insignificant causes.[1] As a result, applications developers spend disproportionate time and effort addressing the interfaces, per se, rather than the application itself.

Standards support the passing of information between the various elements of a manufacturing enterprise's information system. For example, they facilitate connections from computer-aided design (CAD) to computer-augmented process planning to computer-aided manufacturing (CAM). Standardization provides the benefits of common systems (cost savings, improved integration ("plug and play" equipment and systems) and information flow, and so on).

The vision for 21st-century manufacturing presumes that interconnecting manufacturing applications will be as simple as connecting household appliances to a power grid—one need only know how to run the application (equivalent to using a microwave oven) and manage the interface (plug it in and press a few buttons). This ease of interconnection and interoperation extends from devices found on the factory floor to applications connecting the factory to the product design facility to applications connecting an enterprise to its suppliers and customers. An "application socket" for manufacturing would benefit both equipment vendors and customers and enhance factory performance.

The desirability of common standards notwithstanding, businesses have largely been unable to agree on protocols or standards of communication for different kinds of requests. One reason for this failure has been inadequate technology (e.g., the use of primitive fixed-format data representational formats) and excessive complexity. But a more important reason is that attempts to reach a common technical understanding have been hampered by epistemological differences reflected in the unstated assumptions of each participant in a technical discussion. Social factors relevant to the adoption of standards are discussed in Chapter 7.

Research is needed to develop better manufacturing architecture, standards, and interfaces, including research to develop standard equipment control architectures and generic functionality within the architectures, to support general manufacturing information standards, and to lower the cost of more open, less proprietary approaches. Especially desirable would be architectures whose standards accommodate some upgrade capability, so that technology vendors could worry less about premature freezing of technology and the locking out of poten-

---

[1] Wysk (1992) identified as an obstacle to implementing computer-integrated manufacturing the difficulty of integrating information system components (i.e., hardware and software) with information (both internal and external) into a smoothly running system.

tial competitive advantages, while customers could worry less about intrinsic obsolescence.

The importance of standards for manufacturing interconnections raises the question of where technology research enters into the development of standards; the answer is not clear. For example, efforts to develop PDES/STEP (product data exchange using the standard for the exchange of product model data) are not research per se, but the kind of problem that PDES is intended to solve may require research.[2]

Because of the importance of international standards in the global marketplace, strong participation of U.S. interests in standards making is valuable, and greater involvement of U.S. academic researchers (who are less involved than researchers in other countries) in standards making may strengthen the U.S. technology base in manufacturing. Research is a foundation for the consensus building inherent in standardization, and researchers (from industry, academia, and government) should bring their insights to bear on that process.

INTEGRATION

Redesign of Business Practices

Today, workers talking on the telephone, typing on keyboards, obtaining parts, scheduling factory operations, and setting up delivery schedules contribute much more to manufacturing costs than do the workers performing "touch labor."[3] Lessons learned in both service and manufacturing industries attest to the value of systematic redesign of business practices and to the key role played by information technology and infrastructure in such reengineering.[4]

In particular, workers in a factory may be working very hard on an inappropriate set of tasks. They may be doing things that are not necessary, doing them inefficiently or even redundantly, or doing their work in such a way that it is not usable or accessible elsewhere. Information technology can be used to help identify such problems. For example, when modeling techniques are used and

---

[2] The PDES efforts are not standards making in the traditional sense. Like many other activities that have moved into the standards arena, they involve significant levels of development activity as opposed to simple harmonization of approaches. This is true in many other areas of high-technology standardization, such as the unified approach to conceptual data modeling.

[3] For example, Miller and Vollmann (1985) assert that "overhead costs as a percentage of value added in U.S. industries continue to increase and have become a major concern of manufacturing managers" and that "the growing implementation of factory automation increases overhead costs as a percentage of value added." They point out that "most overhead costs are created by transactions, not physical products. Such hidden factory costs are generated by logistical, balancing, quality, and change transactions."

[4] See Chapter 6 in CSTB (1994a). For a discussion of business process reengineering, see Hammer and Champy (1993).

the model is compared with an actual process, steps that do not add value can often be identified. Such information is needed if management is to support a process of continuous change and improvement. Once such problems are identified, information technology can often play a key role in restructuring business processes for greater effectiveness and efficiency. IT can help to reduce the number of tasks involved in an activity, change the nature and sequence of tasks, and reduce the total time involved in a business process

Intra-enterprise Integration

Within a manufacturing company, activities such as design, production, marketing, finance, sales, use of human resources, and distribution have historically been linked to each other on the basis of issues such as ease of management and physical proximity. For example, using the manual and semiautomated manufacturing information systems of the past, integration was achieved through interactions between people where knowledge, data, and project status were shared through formal conversation in meetings and reviews and informal conversation in the hallways.

Today, electronic mail, shared and remotely accessible databases, "groupware," and other forms of electronic communication present new options for centralizing and dispersing activities and also for supporting cross-functional teams and processes. For example, enhanced communications options tend to be associated with greater decentralization, obviating the need for all tasks to flow through an office or headquarters. Electronic communications enable a much higher degree of geographic independence. While communication has always been fundamental to large organizations with distributed establishments, IT supports ever greater geographic dispersion of personnel, allowing firms to reap the benefits of lower-cost production labor or specialized expertise associated with specific geographic regions.

What has stood in the way of a greater degree of intra-enterprise integration? There are many causes, but one major reason is that the technical integration of various applications into a smoothly running system has been and continues to be quite problematic. Commercial software from one source does not integrate well with software from other commercial sources or with internally developed software, revisions come out slowly and hence frustrate users who expect a rapid response, and so on. System-level issues have received relatively little attention, both because of their difficulty and because they relate to problems that fall between the offerings of individual vendors. Customizing of general applications to specific customer needs and integrating applications and technologies from a variety of sources make development of a robust manufacturing system a formidable task. Such problems have plagued all kinds of organizations, but the combination of physical and information-based work in manufacturing adds further complications.

For example, several manufacturing companies have invested many thousands of person-hours in developing and integrating factory information systems, a disproportionately large investment in comparison with their investment in other parts of the factory system. This imbalance is especially serious considering that these companies have often purchased "finished" software application systems from commercial sources, which when measured against the benefits received have proved to be less than satisfying. Since large-scale systems for manufacturing are often one of a kind, they present special problems because they are developed by teams with relatively limited experience with the kind of system needed, and they generally entail very high life-cycle costs.

A second illustration is the integration problems associated with CAD tools for mechanical products. Each vendor of such tools has its own database design, and exchange of models between different CAD programs is possible only for relatively simple constructs, despite the existence of a standard called the initial graphics exchange specification (IGES). Each CAD tool has its own unique interface that may take months to learn, and each comes with a different set of software tools for manufacturing support that may provide functions such as numerically controlled machining, sheet metal bending, or electrical wire routing.

Such incompatibilities are found throughout various manufacturing applications, and every phase of the design and manufacturing process requires its own distinct computer model and data representation. The representation of a part in a database may appear differently for a machining operation than it would for an inspection operation. The design representation necessary to support an engineering drawing is different from the design representation necessary to support a finite element analysis, and in most cases there are no automatic algorithms available to convert from one representation to another. For data other than data on the shapes of parts, few good computer representations exist at all. Even when the interfaces are perfect, a small design change requires a tedious and labor-intensive process of propagating the changes through all the CAD tools and representations. Box 6.1 elaborates on the fundamental knowledge gaps that underlie difficulties in integrating applications.

Databases and database management systems, equipment control systems, communications systems, and design support tools all present a multifaceted challenge to achieving interoperability. Meeting that challenge requires a deep understanding of underlying data constructs, the operations that constitute the relevant manipulations, and the assumptions behind various operations and calculations.[5] While now readily available in certain domains (e.g., the representa-

---

[5] For example, Defense Department officials tell a Gulf War story regarding an inquiry about the number of MREs (Meals, Ready to Eat) available. The answer that came back was "–38,000." This ridiculous answer was the result of incompatibilities between two databases: one defined an "MRE" as a single individual package (one meal for one soldier), while the other defined an "MRE" as a pallet of 500 individual packages.

**BOX 6.1 Knowledge Gaps Relevant to
Intra-enterprise Integration**

• The science and engineering of many phenomena underlying today's products are not well understood. Examples include the behavior of composite materials, failure modes of complex computer systems, turbulent fluid flow, failure modes such as crack growth and fatigue, and material and chemical incompatibilities.

• The models that do exist capture one phenomenon, behavior, or mode of energy storage at a time, while most complex systems exhibit multiple modes at once. Integration of single-mode models is also problematic.

• While the main dimensions of parts and parameter values of an electrical or chemical nature can be specified with reasonable certainty, less is known about how much variation about those nominal values is permissible. The current means of setting tolerances are historical or very conservative. An example is that materials specifications are often stated without indication of variances. The result is that costs are raised and materials and energy are wasted, both in production and in the over-engineering of products that will perform even if tolerances happen to reinforce rather than cancel each other.

• The most efficient ways to carry out design processes to address all the interacting variables are not known. Right now an iterative and "concurrent" process is used, relying heavily on human memory and face-to-face communication.

• It is not known how to construct and populate a database that will contain in an easily accessible form all the information that the designers of modern products will need.

• It has been difficult to develop standards for data interchange to which all vendors will adhere. Indeed, a vendor seeking greater differentiation between its products and those of other vendors may well choose a proprietary data format or representation. At the same time, customers may not realize the waste and overhead incurred in retraining their personnel to use many different software packages.

• Finally, we do not know how best to link people to a rising tide of increasingly complex information. For example, information systems must be able to provide information at appropriate levels of abstraction: a product designer may not need to know the details of how a partially finished part is held in a jig, although a production manager might. Information displayed for different individuals in the manufacturing system should reflect their different needs. The term "information ergonomics" has been coined to capture this challenge.

---

tion of integer and real numbers), such an understanding is sorely lacking in many of the domains that manufacturing touches.

Commercial interests play a key role as well. Vendors have often chosen to develop products with proprietary interfaces and data representations. Suppliers of manufacturing equipment anxious to "differentiate" their products have obscured generic elements in their control computers, thwarting change and raising the cost of acquisition, operation, and maintenance. Such practices have tended to sacrifice interoperability to the desire to capture a set of customers. Not only do such proprietary designs inhibit communication with competitors' products, but they also block communication with adjacent machines and with older capital equipment and networks previously installed within a single plant. Communication is further confounded by network architectures that typically channel communication into hierarchies that mimic the power structure of an organization, inhibiting or complicating the exchange of simple messages between peers.

The result has been that information systems products from one vendor are not designed to be (and sometimes cannot be) integrated with similar products from a different vendor. (In some cases, even products from a single vendor are incompatible with each other.) Interfaces between these systems and equipment are often crude, causing many problems in their use and, according to many manufacturing executives, discrediting the entire information technology effort. Representations for product designs, process plans, and factory resources are quite often company- or even site-specific.[6]

Architectures (functional, control, communication, and management) also vary excessively. Furthermore, these architectures are difficult to describe and model. Defining architectural linkages and creating integration "hooks" for sensors, machine control, process planning, scheduling of maintenance and repair, and machine interfaces are as critical as creating faster and better devices for each of these areas. To a large degree, the presence of such hooks would give users the flexibility to assemble their own solutions from off-the-shelf software and hardware components.

The top-level goal is the effective and rapid exchange of data and knowledge between all operational systems and subsystems,[7] although such a goal applied to research systems might well freeze technologies prematurely and inhibit improvement. Achieving it will require common languages, operating systems, and networks. Beyond common formats for data representation, a common understanding of data, that is, a common semantics, is also needed for complete knowledge interchange. Three important areas of research on the interconnection of applications include the following:

1. Research is needed on organizing principles and architectures for connecting different network-based applications into a seamless environment. Such connection is necessary, for example, to link flexible manufacturing

---

[6] Internally developed integrated information technology solutions are always possible in principle. But as a practical matter, only the very largest firms have the resources to develop even partial information technology solutions in-house, and essentially no firm can do it all alone.

[7] The need for commonality and interoperability among manufacturing applications of information technology is in some ways analogous to the problems faced by European engineers 250 years ago as they struggled to integrate mechanical systems that utilized a puzzling array of hole sizes, bearing sizes, and thread sizes. The crucial difference is that standardization of the latter took place over decades, while today U.S. manufacturing cannot afford to wait decades for these standards to arrive.

cells to a plant's scheduling function and to link the scheduling function to the enterprise order, delivery, and financial systems. Enterprise integration also implies a need for research to enable the automatic interpretation of the type of transaction being executed, the routing of the message to the right location for processing, and the processing that must occur when the message for the transaction reaches the correct system.

2. Research is needed to support communications networks that can support the real-time control of manufacturing processes (e.g., a cutting tool shaping a piece of metal). Such networks transmit control messages, each of which must be received and acted on within some critical time, Δt. While the process being controlled should be able to tolerate some variation in Δt, the network must be able to guarantee delivery of that message unconditionally within that time. Many common protocol suites such as today's TCP/IP or Open Systems Interconnection deliver messages on a "best-effort" basis and cannot reliably meet such delivery-time requirements, while other protocols assure timely delivery by reserving communications bandwidth, usually on private lines and usually through restrictions on the type and number of messages carried. Since there is increasing demand for the ability to transmit control messages on existing open communications networks, research is needed to formulate the principles for construction and operation of networks that support time-critical message delivery in a context of interconnecting, multipurpose networks. If possible, such systems should be compatible and interoperable with other time-critical and non-time-critical applications and communications; in this event, they could be made compatible with a wide range of TCP/IP products.

3. Research is needed to develop services tailored for human- and machine-based information and resource browsing and searching, using knowledge-based assistance agents for semantic interpretations, translations, and relationships. The availability of such services will be essential to ensuring greater information infrastructure performance and efficiency. The higher-level network directory services of today will not meet the requirements of near-future applications.[8]

Inter-enterprise Integration

The notion of a supply chain (Figure 6.1) will be integral to manufacturing for the foreseeable future, and suppliers will be passing partially finished components or materials to those closer to the point of final assembly. Increasingly,

---

[8] Some relevant work has been undertaken in the context of the X.500 directory standard, and further work on implementing that standard is necessary.

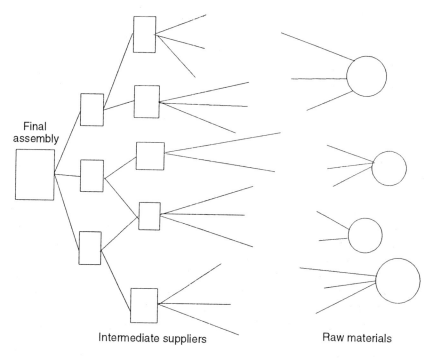

Final
assembly

Intermediate suppliers   Raw materials

FIGURE 6.1   The supply chain, a basic element of manufacturing that will become
increasingly information intensive.

along with the movement of goods, information must move up and down the
supply chain. Information technology can facilitate the passage of information
within a manufacturing enterprise, but as importantly, it can enable different
organizational structures and relationships among various elements in the supply
chain, and networked infrastructures can extend well beyond a firm's boundaries.
IT can be used to link value chains across firms and shift work involved in
various functions between organizations. Already, electronic data interchange
(EDI) connects companies to their suppliers, shippers, buyers, and even their
banks. Providing a shared body of information and a common set of standardized
forms, EDI transforms the ordering, invoicing, shipping, tracking, and customer
service functions, among others, for a growing number of enterprises. Through
these electronic linkages, IT enables companies to emphasize their core competencies by "outsourcing" activities that are mostly common across industries
(e.g., payroll, purchasing, or accounting) to other providers that can perform
these activities more efficiently. In addition, electronically linked firms can
group and regroup as the need arises in alliances of convenience.
   At present, much of the material-related cost of production stems from the

purchasing, packaging, shipping, unpacking, and moving around of component materials. Better communication between supplier and customer enterprises would reduce such costs; although not new, EDI and other tools to support streamlined processes are obvious points of departure and candidates for supporting technology that will eventually culminate in the electronic communication of financial information and even electronic funds transfers, involving corporate-level systems and systems at third-party financial institutions.

Mechanisms and systems are required to support information session management. These include the means for connecting to and coordinating the delivery of information between multiple sources using multiple streams, as well as the mechanisms for collaboration using this information. In addition, tools and techniques are needed to facilitate real-time planning through the entire supply chain; in certain factories, this is already a partially achieved goal, but for a variety of reasons the techniques are not widely applicable across all manufacturers. Intelligent agents (computer automata) may be able to perform most of the hand shaking between customers and suppliers without requiring human intervention; if realized, these agents might provide human users with new, flexible, easy-to-use human-oriented protocols and standards that are as obvious as the current paper-based solutions. This level of analysis is generally impossible today; given the typical operation of manufacturing resource planning systems, it can take weeks. Research will be required in terms of software architectures, protocols for communication, and access and security mechanisms (see "Dependable Computing Systems" below). These needs are above and beyond the transaction-based processing and management that will clearly be required.

NON-MANUFACTURING-SPECIFIC RESEARCH

Given that manufacturing operations are typically large and complex, it is not surprising that the software for controlling and managing such operations is also large and complex. Research in many traditional areas of computer science, including information retrieval, software engineering, and reliable computing, as well as in collaboration technology, will be relevant to manufacturing. Work by computer scientists in these areas, but framed in a manufacturing context, will both enrich computer science and result in important benefits to manufacturing.[9]

---

[9] An approach calling for research work at the intersection of computer science and engineering (CS&E) and other problem domains was contained in the CSTB report *Computing the Future* (CSTB, 1992). In particular, this report called for academic CS&E to increase its contact and intellectual interchange with other disciplines (of which manufacturing could be one); to increase traffic in CS&E-related knowledge and problems among academia, industry, and society at large; and to enhance the cross-fertilization of ideas in CS&E between theoretical underpinnings and experimental experience.

Information Retrieval Systems

Manufacturing applications are data-intensive. Enormous amounts of data are needed to support manufacturing activities (e.g., to represent the millions of parts on a Boeing 777 or the millions of transactions that occur on a factory floor), and the relationships among the data are complex, sophisticated, and changeable. For example, a very large scale integrated chip component has a different description for each activity; functional, geometric, behavioral, temporal, and other views of the part must be supported. Any changes in one view must ripple through all of the other views and extend to invalidate or update usage of the component throughout a design process. The movement toward multimedia databases and applications compounds the problems presented by simple text or numerical databases.

Modern information storage and retrieval systems are designed to handle all aspects of data and information management except generating or collecting the data. These systems serve as the primary interface to data for the user through powerful query languages, for bulk storage devices through the file system, for the rest of the computer through the operating system, and for the rest of the world through communications networks. In each of these areas, advances have been rapid, but progress has only generated more ambitious goals. Information retrieval for manufacturing will stress existing technology and will require a major research effort. The following three problems illustrate the type of work that is needed:

1. *Incorporation of modeling and prototyping functions into a manufacturing database management system.* The language used for data manipulation within the system should contain the basic syntax and semantics that describe products and processes; in other words, the product and process description language should be a sublanguage of the data manipulation language. However, current data manipulation languages were developed to handle financial and other types of data that can easily be cast into a relational database model. Manufacturing data, especially data related to geometry, are instead associative. Consequently, using current data manipulation languages for manufacturing data is counter-intuitive and difficult. Research is needed to develop the next generation of data manipulation languages, perhaps based on the object-oriented data model.

2. *Maintenance of data consistency and integrity in a rapidly changing database that may be distributed.* In the design phase, the database must serve not only as the primary repository of ongoing work but also as the principal medium of communication among the participants (who may be using simulation, analysis, and design tools, as well as manufacturing execution systems); these participants may also be geographically distributed

across many time zones. Any disparity in the copies of the database seen by different users as changes are made will wreak havoc. In the processing phase, consistency is essential if real-time control of the process is to be achieved. A real-time production schedule is realistic only if the data about equipment status and work location are correct when the schedule is generated. Here the performance (i.e., speed) issue that is always present in database consistency becomes particularly difficult to resolve. Propagating changes throughout the entire system imposes additional constraints on real-time performance.

3. *Information browsing and searching in a distributed manufacturing environment.* Vendors and suppliers connected through the National Information Infrastructure will want to make information about their products available electronically to customers that may not know about them. Design and production engineers will need to browse various information archives in search of a part or a process that they may need. Thus, while network tools such as gophers, MOSAIC, and Wide-Area Information Search systems are promising starts to the general problem, research is needed to develop more advanced automated tools that accept information on needed parts or processes as input, search or make appropriate inquiries at appropriate sites on the NII, and return to the user a list of those sites and the information found there. Such an effort would also depend on new indexing schemes and data representations that would allow semantically driven searches of both text and nontext information.

Software Engineering

In view of the experiences of the major manufacturers represented by members of the committee and those who briefed the committee, the current state of the art in software engineering is barely adequate to meet current manufacturing needs. The software problem has a direct impact on a manufacturer's ability to respond quickly to a changing marketplace with new products. The reason is that new product development often calls for rapid responses and agile manufacturing, which themselves require advanced information technologies, open systems concepts, and component reuse.

The following list describes some of the most important areas of software engineering research for manufacturing:

• *Tools to support the software engineering process in manufacturing.* Automating complex, intelligent manufacturing systems requires both solid engineering life-cycle approaches and supporting tools. Unfortunately, in many cases, tool development lags significantly behind the use of new methodologies, as in the case of tools to support the recent component-based architecture life-cycle approaches. Companies using such approaches for component development and reuse are suffering from the lack of useful tools for system analysis, requirements management, design, configuration and integration, and simulation. Until such tools become available, the building of large, complex manufacturing systems will remain a costly exercise. For example, more general tools are needed that are not constrained by the limitations of specific programming languages, as well as tools that have aspects of knowledge-based collaboration software.

• *Programming paradigms that enable manufacturing personnel, rather than software development experts, to develop and change application systems for the shop floor or the design laboratory.* This capability will require better methods of developing software and better human-machine interfaces to enable domain-specific software specification. Object-oriented programming is a particular paradigm of interest.

• *Support for faster development of easier-to-use and more effective systems.* Simplification of designs, operation, and maintenance is desired, as are increased predictability of systems, self-healing systems, and system extensibility. Research is needed in system optimization and enhanced system operation and maintenance; better capabilities for rapid prototyping are a particular need. Visualization and human-computer interaction techniques will be key.

• *Accommodation of legacy systems.* Manufacturers have a great deal of investment that they will be loath to give up even when new and improved technology is available. Thus, research on how to facilitate transitions to new technology is necessary. One direction for such research is the development of techniques for encapsulation of legacy systems and of mediator support.

• *Tools for systems analysis.* Although there are many different analysis methodologies, such as responsibility-driven, data-driven, and activity-based methodologies, system analysis remains more an art than a science. How do we compare types of analyses? Are some techniques simply better than others, or do different types of system development or applications dictate a choice? Are the techniques sufficient, or are new ones needed? Research is needed to help answer these questions.

• *Better metrics.* A number of metrics have been developed for the testing phase of the engineering process to measure dimensions of system performance such as speed and timing. There are no metrics that specify how well a problem in manufacturing is understood and formulated. What are the metrics for analysis, and how do we measure our successes and failures? When do we know that analysis is complete or at least good enough? Can these questions be addressed with metrics? Given the frequent need for rapid system development, metrics specifying temporal aspects of analysis should be considered. Are there temporal metrics for analysis techniques (e.g., state-transition diagrams, object models, event-stimulus diagrams) that could be related to system characteristics?

• *Software reuse.* The ability to reuse software and system artifacts and results from associated analyses in the development of new systems or applications would leverage previous investments in expertise, effort, money, and time. Examples of software artifacts from object-oriented analysis and design include object models (including representations of data and behavior) and object interface definitions. One type of research supporting reusability could lead to reference architectures that cut across manufacturing domains. For example, there could be families of reference architectures for continuous processes and for discrete processes. Goals would include minimizing the number of reference architectures for which third-party suppliers must develop software or systems and increasing the ability to leverage previous expertise across projects.

• *Better representations of spatial and temporal dimensions in software.* Software representation currently focuses on declarative knowledge. Research is needed to achieve representations that depict both spatial and temporal aspects of associated data, enriching a system's repository of knowledge and facilitating the visualization, design, specification, monitoring, and analysis of manufacturing processes. Multimedia visualization of these new representations may include "walking through" models or providing simulations of system behaviors in one window while the activated models employed are presented in another window. Such elements would contribute to the realization of virtual factories, which involve enhanced modeling and simulation of processes and products. New programming interfaces that capture the spatial and physical abstractions of manufacturing are essential to allow end users to program their own processes and work flows.

Dependable Computing Systems

Manufacturing plants require continuous operation, creating a need for dependable computing systems. Better technology to support "hot swaps" of software (i.e., changing software without removing the system from operation), continuous availability of on-line services, and fault-tolerant hardware and software are among the technologies needed for dependable manufacturing systems.

Manufacturing provides an application arena for a wide variety of research relating to increased system security and trustworthiness, including access control and authentication. The need to determine and verify the user of a computer system or network is becoming increasingly important as a way of preserving system and data integrity, ensuring that sensitive data and systems are accessed only by those who are authorized, and ensuring the highest levels of system availability. These concerns affect both intra-enterprise and inter-enterprise communications.[10] The manufacturing environment, especially the factory environment, calls for economical and robust technology to address these needs.

Finally, as manufacturing tools become more autonomous, it becomes imperative to ensure the correctness of controlling software, as control software with errors may be a major contributor to factory downtime due to physical damage.

Collaboration Technology and Computer-supported Cooperative Work

The trend toward organizing workers of all kinds into teams with significant levels of decision-making authority gives rise to a need for technology to support collaborative activity.[11] For example, intelligent systems are needed to support collaborative efforts in the design of complex products; they can also facilitate collaboration among factory and other, nonproduction personnel. Research needs include information technology to support empowered work teams of various kinds of personnel and tools for total quality management.

Research relating to technical tools for computer-supported cooperative work (including software, user interfaces, and supporting hardware) should be complemented by research examining relevant aspects of human behavior, education and training requirements, and other similar issues, to ensure both that optimal tools are developed and that they can be used easily.

---

[10] Electronic contract documents and order qualifications, which are envisaged as part of our future enterprises, will require methods for identification, authentication, and non-repudiation.

[11] Aside from the emerging, largely message-based groupware products, today's tools are aimed at supporting individual professional performers rather than collaborating teams.

7

Organizational and Societal Infrastructure

Chapters 3 through 6 discuss a research agenda for information technology related to manufacturing. If this research agenda is fully pursued, the scope of what is achievable will be greatly expanded. However, an equally important challenge is understanding how manufacturing enterprises can actually make use of information technology; even today, much technology remains unused.

Put differently, the effective use of information technology (IT) in manufacturing depends on more than technology research. It also depends on success in (1) understanding the concerns of manufacturing decision makers; (2) motivating effective collaboration and technology transfer between industry and academia; (3) motivating individuals within manufacturing enterprises to implement information technology; (4) developing open standards and appropriate metrics of performance; and (5) ensuring through education and training that the skill base in manufacturing adapts to the new types of tools, techniques, and organizational structures made possible by information technology and to other "best practices" and "benchmarks" not necessarily associated with information technology. Also needed will be execution of global business strategies that protect high-value-added jobs for U.S. manufacturing facilities. In some instances, these essentially practical recommendations contain their own requirements for research, particularly social science research. The challenge for the nation is both to improve on existing capabilities (the focus of the technology research agenda described in Chapters 3 through 6) and to enhance the potential for manufacturing firms to actually use advanced information technology successfully (the focus of this chapter).

TARGETING THE DECISION MAKER

Planned change is driven by decision makers. Of course, these decision makers may not necessarily reside in the corporate executive hierarchy; nearly all employees in a company have some decision-making authority. However, people at different levels of a hierarchy have different concerns; as a result, they look to new technologies to answer different questions, as suggested in Figure 7.1. Technology researchers who wish their innovations to be adopted must craft their research in a way that the benefits of the research are highlighted to match the concerns of the decision makers who will decide on their technologies as well as those who will use them. Focusing on the technology alone is rarely sufficient to persuade a decision maker to adopt a particular solution. Unless the use of information technology demonstrably provides substantial payoffs for the end user, the end user will not take the trouble to learn and adapt to the technology, and that will guarantee that the new technology will not be used.

A good example of technology fitting the needs of particular users is evident in the rapid diffusion of personal computers into the workplace. When personal computers became available for a few thousand dollars apiece, they could be deployed to address the problems of decision makers who controlled budgets on that scale without the need for action at higher levels of authority. However, the purpose of introducing personal computers was not to proliferate them or the technology they represent; their use is justified because they can solve real factory problems with a demonstrated dollar benefit.

MOTIVATING TECHNOLOGY TRANSFER
AND ACADEMIC-INDUSTRIAL INTERACTION

No matter how good are the ideas and advances developed by academic research groups, they are useless in a manufacturing context unless transferred to industry. Effective transfer requires a stronger relationship between academia (or other sources of new technology) and industry.[1] Although most of the important recent developments in design and manufacturing have come from industry, manufacturing firms and their information technology system vendors need the knowledge and expertise of the academic research community, whose role is to pursue longer-term research that may not have obvious immediate payoffs. This

---

[1] For example, new design methods developed in academia are especially hard to transfer to industry because industry normally gets such methods from computer-aided design (CAD) vendors. So the transfer takes two steps (from developers to CAD vendors and then from CAD vendors to their customers). However, CAD companies are typically small and very limited in their resources, and so they do not take risks on new research ideas. Instead, they take their cues from their major customers. Consequently, a technology transfer strategy for design methods must enlist the support of potential users as well as equip the CAD vendors.

Who	Where	What, Why, and How

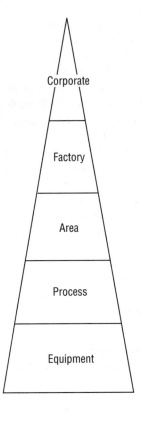

Chief executive officer
Chief financial officer
Chief operating officer
Worldwide marketing
  and sales executive
Chief technologist

What is my production
  capacity?
Who gets what when?
Are we profitable?
What products do we make
  next year?
Do we have the right skill
  mix?

Factory manager
Production manager
Engineering manager
Information systems
  manager
Human resources
  manager

How is my factory doing?
What is the cost?
What are the critical issues?
How are the suppliers?
Is the system responsive?

Area supervisors
Systems operators
Production schedulers
Yield managers

How is my area doing?
Is the information current?
Are we on target?
What is the area yield?
Is the material on time?

Shift supervisors
Process engineers
Equipment managers
Quality engineers
Information engineers

Is my shift operating well?
What is today's schedule?
Is the process in control?
Is the product quality high?
Systems running smoothly?

Equipment operators
Processing technicians
Maintenance personnel
Information
  technologists

What is the equipment status?
How do I fix this process?
Where is the material?
What lot is run next?
Do we have sufficient spares?

To enable a correct response to the higher-level questions,
answers to the lower-level questions must be correct.

FIGURE 7.1 Questions asked at different levels in the manufacturing hierarchy to which information technology can be expected to be responsive.

is not a new problem, but it is particularly acute in the area of manufacturing technology—and resolving it is more important now than ever before, especially since organizations long known for their ability to facilitate the transfer of basic research to practical implementation (e.g., Bell Laboratories, Fairchild R&D, IBM's T.J. Watson Research Center laboratories) are being cut back, reoriented, or even eliminated.

The success of computer-aided design (CAD) programs for electronic design in the marketplace is instructive and provides a good model for university-industry relationships in other manufacturing domains. CAD programs for electronic design have been developed over the last 20 years, primarily in universities and industrial laboratories, often aided by governments grants (mostly from the National Science Foundation (NSF) and the Advanced Research Projects Agency (ARPA)). The transfer of these technologies from laboratories to industry is due to a robust and aggressive set of private-sector vendors, whose role has been largely to implement and package research results for general consumption. Electronic design analysis research is still going strong, primarily in raising the level of design abstraction and in improving the capabilities of existing algorithms. Development and deployment of field-programmable gate arrays are starting to make it possible to build early hardware prototypes, rather than relying on software simulations. Again, private-sector vendors (Xilinx, Actel, Altera for chips, Quickturn for systems) are marketing the basics needed for such prototyping.

Some of the seeds of the problem of transferring technology for manufacturing lie in the culture of universities. Unfortunately, manufacturing is dismissed in many academic circles as a shrinking portion of the gross national product symbolized by unexciting smokestack factories.[2] Inducing academics to work on the inherently interdisciplinary and multidisciplinary problems that characterize manufacturing problems is difficult when tenure in universities often depends on evidence of the faculty member's ability to be successful as an independent researcher gaining peer recognition; teamwork and industry collaboration are not regarded as important contributions. Industry recognizes this fact by forming design teams and finding ways to reward all the participants. In academia, many forces are starting to motivate collaborative research, which is presenting new challenges and opportunities to administrators; these include, for example, interdisciplinary manufacturing research centers (CSTB, 1993).

Industry can help by articulating to academia the basic intellectual issues of manufacturing, not only to help guide research (e.g., by providing data and problems to academic researchers) but also to make those issues recognizable to people in traditional disciplines. Strong statements from industry would help to impress on manufacturing faculty (many of whom have limited expertise in infor-

---

[2] For a perspective on the larger problem of nurturing academic work related to system development, see CSTB (1994b).

mation technology) the need for industry involvement and the increasing importance of information technology to manufacturing. Finally, if industry were
widely regarded as being open and accepting of new ideas, academic researchers
might have more incentive to work on industry's problems.

A variety of mechanisms could foster better academic-industrial interaction
relating to manufacturing. A fellowship or sabbatical program, for example,
could place academics in factories and factory people in academia for periods of
about a year. (Shorter periods have proven less satisfactory; periods of about a
year allow a researcher to better understand the special qualities and inherent
problems of the manufacturing environment, to explore its different operations
and dimensions (e.g., by rotating through different units), and to contribute substantially to addressing a problem.[3]) Historically, this concept has been thwarted
by tenure pressures that militate against long absences from the university by
researchers during the period in which they are forming their research programs.
To succeed now, mechanisms are needed to ensure that researchers are not penal-
ized for this kind of investment of time and effort.

The committee is particularly attracted to the concept of a "teaching factory." The study of manufacturing requires a strong laboratory experience, in
which difficult concepts and poorly understood interactions can be demonstrated
and learning can be reinforced, and a teaching factory is designed to provide this
laboratory. A teaching factory would gather together the necessary processing
and control equipment (including tools, fixtures, machines, computers, and interfaces) and organize these elements so that a product or family of products could
be produced. However, the teaching factory would also be organized to demonstrate and make manifest a variety of activities and principles, including the
specifics of processing, control, system and product design, associated management, and their interactions. A teaching factory would serve as a testbed that
could help researchers see if their ideas are valid. Setting aside portions of real
factories for such experiments is usually too costly to be considered. Yet testbeds
must be realistic enough to enable researchers not only to find out if their ideas
work but also to learn the constraints of real manufacturing environments and
discover opportunities and research problems.[4] Research to develop innovative
and economic means of creating teaching factories is necessary to prepare manu-

---

[3] Industry could reciprocate by funding further education for its workers. One example of a useful
exchange is the Massachusetts Institute of Technology's Leaders for Manufacturing program, which
links engineering and business school faculty with industrial representatives and involves master
theses developed by students "on location" in manufacturing environments.

[4] The Metal Oxide Semiconductor Implementation Service (MOSIS) provides a facility to experiment with innovative chip designs; a researcher provides a chip design, and the MOSIS service
returns (in a very short time) a chip fabricated to the design's specifications. The concept of a
"mechanical MOSIS" service has been proposed in the past, and it continues to be attractive as a
vehicle for rapid testing of interesting manufacturing ideas.

facturing specialists for the factory of the future as well as to develop deep understanding of specific processes.

One variation on the concept of a teaching factory—the extension of the manufacturing enterprise into academic institutions—would be made possible by a broad, enhanced national information infrastructure (NII) such as that envisioned as enabling greater inter-enterprise integration and the virtual factory concept (see "Inter-enterprise Integration" in Chapter 6). Efforts to develop a capability for remote access should specifically address problems related to electronic connectivity, the transfer of technically complex principles, and shared access (perhaps remotely) to expensive manufacturing equipment and critical information or knowledge.

Any program that provides major funding for academic research in manufacturing will likely increase academia's interest in manufacturing; of course, that is an objective of the federal Advanced Manufacturing Technology initiative. Another mechanism for boosting the prestige of manufacturing would be an NSF postdoctoral program for manufacturing fellows. Matching grants would leverage limited federal funds with industrial resources, using industrial support to signal areas of particular interest to industry. Similarly, funding could be directed to appropriate university-industry partnerships that are launched with explicit mechanisms for technology transfer.

Finally, consortia may be necessary to undertake meaningful manufacturing research, a trend recognized in a number of federal programs.[5] The issue of manufacturing is bigger than any company, indeed bigger than any industry. Consortia may be the way to carry out expensive research, especially in areas that transcend specific products. One may question the relevance of, for example, the possible collaboration between an aircraft company and a semiconductor chip

---

[5] For example, the Intelligent Manufacturing Systems (IMS) program is an international program for cooperative research and development in advanced manufacturing that involves the United States, Japan, Canada, Australia, the European Community, and the European Free Trade Association. It is intended to bring together large and small manufacturing companies, research and academic institutions, and public authorities in a structure that facilitates the sharing of intellectual property. In the United States, the IMS program is run out of the office of the assistant secretary for technology policy in the Department of Commerce. This program is not aimed at any particular research topic or direction for attention, though certainly the application of information technology to manufacturing does fall under its umbrella.

A second example is the technology deployment activity of ARPA's Technology Reinvestment Project that focuses on manufacturing extension centers (MECs). An MEC is an organization that works with small manufacturers to assist them in using technology (including information technology) to improve performance and productivity. MECs are intended to be consortia consisting of institutions of higher education, nonprofit organizations, or government entities in the service of specified clientele.

Other consortia are described in *Putting the Information Infrastructure to Work* (NIST, 1994, pp. 7-23).

company. However, it turns out that there may be many possible benefits of collaboration. Boeing has established a sophisticated design program for its new line of aircraft, and the supplier-customer relationship established to share designs of parts can benefit a chip maker also trying to share design topics. Both companies share concerns about environmental issues, team development, and technology to share information. Interindustry collaboration may indeed provide a rapid path to the introduction of new technology.

The development of the NII is sure to have a significant influence on industry collaboration. Sharing of information will lead to more collaboration and the availability of high-bandwidth communications networks will facilitate more and better ways to collaborate, as will the availability of an infrastructure involving adoption of standards that support interoperability. As graphic and image-rich information management capabilities become available through the NII, for example, the collaboration between industries and suppliers is very likely to be enhanced.

MOTIVATING INTRODUCTION AND IMPLEMENTATION OF INFORMATION TECHNOLOGY

Understanding Organizational Issues

The introduction of new technologies into an organization almost always affects existing social structures. A factory operation that is as tied to information technology as the vision of Chapter 1 suggests will inevitably have different employment hierarchies, divisions of labor, information flows, and forms of organizing for tasks than those that exist today. To facilitate a smoother integration of new information technology into manufacturing businesses, research is needed in the following areas.

Human Resources

Even in a highly automated and information technology-intensive manufacturing enterprise, human beings will continue to have a critical impact on operations. Important questions regarding the human role must be addressed:

• How should responsibilities be divided between people and information technology? At the extremes, this division may be clear (clearly, people should decide if it is socially responsible for a certain product to be produced, and computers should decide at what moment to turn off a cutting torch). However, intermediate cases are less clear, especially as information technology becomes capable of acting with more autonomy and can assume tasks previously thought to require human input.

• How should goals and incentives be structured to ensure that the aggregate of individuals' behaviors is most supportive of an enterprise's overall objective? Individuals in a hierarchy may well be much more sensitive to the demands of their immediate environment than to the company's overall goals; most individuals in an organization do what is necessary to satisfy their supervisor, and what the organization wants overall is secondary. In motivating the appropriate behavior, a system that rewards individuals on the basis of satisfying the needs of (internal or external) customers may be more efficacious than current approaches, and information technology may well provide the means to make such a reward system more immediate and thus effective (Holland, 1985, pp. 1-7).

• What mechanisms are needed to give individuals the skills they need? This issue implies training or apprenticeship programs and invites exploration of issues such as just-in-time training for small firms that cannot afford to release staff for more traditional educational environments. Older workers who associate high technology with increased job insecurity may be particularly apprehensive about the introduction of IT.

Communications

Traditionally, the content of communications is structured as distinctions between data, information, and knowledge. This continuum can be regarded as one of ever-higher abstraction coupled to increasing levels of contextual awareness and relevance. Intuitively, data belong to "low-level" computers while knowledge is the province of "high-level" human actors. But as computers become capable of assuming higher-level functions, what data, information, and knowledge are relevant to which entities? (For example, a person or a computer might have a single function that requires selected communications content from across an entire facility, or it might deal with the entire communications content regarding a geographically and temporally bounded environment. Under what circumstances is each design more appropriate?) How can the communications stream best be made available to workers and for what purposes? How can the differing requirements of individual agents be taken into account? How can information overload on human beings be detected and corrected?

Organization for Ad Hoc Tasks

For substantial tasks, teams are necessary. As enterprise integration for manufacturing expands to include actors not traditionally included in a manufacturing factory complex (e.g., suppliers and customers), issues of team management and organization will assume added importance. For example, the problems of adjusting organizational structures to permit and support flexibility in design and increased interaction with suppliers and customers as they participate in product development constitute one of the fundamental issues manufacturers are wrestling with today. Moreover, the group dynamics of teams with human and computer actors are very poorly understood. Put differently, what should the partnership between person and machine be like?

Informal Hierarchies of Status

In addition to a formal organizational hierarchy (as might be depicted on an organizational chart), most organizations also have informal hierarchies that reflect the greater status or prestige of certain categories or classes of individuals relative to that of others. In manufacturing companies, product designers are often at the top of the pecking order, even though a successful manufacturing operation requires talent and intelligence in other job functions as well. Such status is consistent with the fact that product designers have a much richer array of computer-based tools to help them in their work than do factory managers or process designers. One consequence of this richer array of tools is that product designers can point to computer-supported decisions whereas others usually must rely on experience and intuition, and it would not be surprising if the preferences of product designers were given more (and perhaps undue) weight as a consequence. Preceding chapters have underscored the fact that knowledge about product design is more advanced than that for process design or factory management, but it is also clear that limited resources are more often invested in accordance with the needs of those of higher prestige and influence. Because of a long-standing history of status differentials in U.S. manufacturing organizations, it may be necessary for management to pay explicit attention to differences in status and to make special attempts to listen to the needs of individuals across an organization so that they can contribute on a more equal basis.

Overcoming History and Managing Risk

The history of attempts to introduce computer-integrated manufacturing (CIM) systems into factories demonstrates that companies have tried to adopt these systems without adequate justification or based on the wrong metrics or parameters; the result has been that these large expenditures of dollars have been poorly justified and often wasted. As a result, many decision makers in manufacturing regard CIM technology as technology searching for problems to solve. In some cases, such a perception may taint even good CIM applications.

Moreover, the gap between manufacturing managers (the customers for manufacturing research) and technology "pushers" (the suppliers of manufacturing research) is often large; technologists at times push technology because it is "neat stuff," whereas managers want and need answers to today's problems. This gap between technology pushers and customers with real problems is significant and is the source of much friction. Many information technology researchers are so immersed in computer technology that they take increasing computerization as a given. These researchers often suffer some hubris about "those people out on the factory floor who cannot write programs or integrate new systems . . . ." Manufacturing managers, including members of this committee, express concern that they already have more technology than they can use. They note that many human obstacles must be overcome before today's, let alone tomorrow's, information technology can be used to benefit manufacturing. Attention to the managerial and cultural preconditions for effective use of information technology is an important element—perhaps the most important element—of the development and application of advanced manufacturing systems (CSTB, 1994a).

The gap between technologists and managers was demonstrated directly to the committee in its interviews with senior manufacturing executives, who spoke in terms of cost, quality, and time to market while the committee spoke in the jargon of information technology. The company executives' view of what was contained in information technology differed strongly from the committee's view; the committee's view of manufacturing was strongly at variance with the executives' view. As one of the interviewees for this report put it, "The managers and the technical experts talk two different languages. The technologists regard the managers as being too tied up in return-on-investment concerns and therefore likely to miss emerging capabilities, while the managers and decision makers regard the technologists as being too uninvolved in real factory problems to understand what the customer really wants and needs." Such discussions demonstrated that both technologists and executives need to make greater efforts to bridge the communications gap before technology can be applied to the real problems faced in manufacturing in companies across the United States.

Manufacturing managers know that realizing the promise of information technologies to help improve factory performance is not easy, and there are many opportunities for mismatched expectations. Expectations may differ between technology supplier and user—indeed, users may have had too little input into the design of new systems. Achieving a solution may take so long that it may come only after the problem has largely disappeared or been fixed by other means, or the conditions have changed so much that the solution is no longer applicable. A new system may not make use of embedded legacy systems that are too expensive to discard or too integral to operations to modify. The skills required may differ significantly from those of the existing work force (see "Education, Training, and Retraining" below).

Thus, the resistance of many senior managers to the introduction of information technology into real manufacturing operations is not unreasonable, and it arises from the fact that many of the new information technologies are untested and unproven in actual manufacturing lines, and in some cases have offered poor results. New systems are not used because they are untested, and they cannot be tested because no one wants to be the guinea pig.

BOX 7.1 Guidelines for Introducing Innovative Information Technology in Real Production Lines

• *Take small steps.* Deliver improvements incrementally, rather than striving to deliver "the solution" all at once. Deliver improvements that can be installed with lower levels of management authority.

• *Lower the risk.* Avoid serious disruption of manufacturing facilities while applications are being tested or while systems are being introduced onto the factory floor.

• *Deliver real solutions to real problems; focus on the customer, not the technology.* Indeed, there is a great deal of good technology already waiting to be used. Develop the customer "pull" in concert with the technology "push."

• *Develop a vision and a set of strategies that have temporal longevity.* Create a system that withstands the ups and downs of yearly budget plans. For example, "lean" manufacturing is still not 100 percent utilized, in spite of the fact that it was started over 30 years ago!

• *Work in teams with customers and suppliers.* Since having a viable supplier base is a key to success, it is important to make suppliers a part of the solution, rather than regard them as part of the problem. Similarly, the known (and perhaps anticipated) needs of customers should be the ultimate focus of installations of information technology.

• *Develop systems that build on legacy systems,* unless it appears that the legacy system is so hopelessly useless that it must be jettisoned. It is hard to quickly change the momentum of manufacturing enterprises; try to make incremental improvements.

• *Where possible, buy off-the-shelf information technology from commercial suppliers.* In general, it is better to use commercial vendors than to undertake custom development projects on one's own. Doing the latter forces one to inherit all the issues that relate to product and application support for the product life cycle.

Another contributor to risk aversion is time pressure. Bringing in new systems consumes valuable production time, which may not easily be recovered. It is easier (and therefore common) for a factory to use an old, perhaps less efficient, system than to suffer "downtime" as the result of installing and testing a new system. Box 7.1 provides some guidelines for introducing new information technologies that are responsive to the difficulties that line managers face in approaching these technologies.

The scope of change inherent in manufacturing systems can confound financial analyses. The financial impact may appear too large to justify for a single factory, even if the new technology appears worthwhile for the enterprise overall. In addition, the benefits of investing in a new system may be hard to quantify, especially if they relate to quality; metrics for improved quality, service, and accessibility are particularly problematic. Research is needed to support more appropriate financial and business justifications and to better define the cost of quality, the cost of insufficient quality, and the trade-offs among quality, time, and money (Box 7.2).[6]

One mechanism that could contribute to better planning for the time and real cost involved in implementing new systems (although not necessarily changing those quantities) would be an extension of the "beta test" concept into manufacturing environments.[7] Beta tests can assess the fidelity of a product to its design specification, augmenting a real production environment with special instrumentation to assess errors or relay feedback to developers. Although pilot plants have been used for years as laboratories to assess production concepts, research is needed to explore how actual production facilities, operating at normal volumes, can be used directly for achieving continuous improvements or innovation. Legitimizing beta test periods would disclose the need for and real costs of a practice to manufacturing management at the proposal stage. The beta test concept could also provide a basis for links to research facilities, with the benefit of enhancing the understanding of what does and does not work in a given type of system or application. Research could help to define the tools, techniques, and practices appropriate to beta testing in manufacturing contexts.

Finally, the development of the NII may, for entirely psychological reasons, stimulate the incorporation of information technology into manufacturing. To the extent that the NII is a medium that ordinary people use every day (e.g., to make purchases, to read newspapers, to order and watch movies, and to transmit pictures, images, and voice and other messages), manufacturing managers will become more familiar with information technology, and the part of managerial resistance to information technology in factories that is attributable to unfamiliarity will likely diminish.

Providing for Technology Demonstrations

In the absence of convincing evidence that the benefits of information technology far outweigh the costs, effects of factory disruption, and time required to do the job, many managers and executives in manufacturing enterprises will be understandably risk-averse. For these individuals, there is a world of difference between a promise and a concrete demonstration of a technology's feasibility and

---

[6] For example, the economic production of customized manufactured products (i.e., production lot sizes of one or a few) may well depend on the adoption of new accounting systems in which the entire life-cycle cost of a product is taken into account; life-cycle costs include transportation costs, inventory costs, obsolescence costs, reengineering and retrofitting costs, and the cost of product recall, if any.

[7] Software developers draw on objective third-party users in beta tests of new products or versions to verify functional capabilities and ensure lack of defects. Explicitly applying the concept and practice of a beta testing period into production launch activities would provide for significant first trials of the custom software typical of manufacturing information systems.

> ## BOX 7.2 Information Technology Deployment and Justification
>
> Information technology was originally used in fairly narrow ways: accounting and clerical functions were automated, personnel records were placed in rudimentary databases, and controllers were developed for individual machines. For these applications the financial justification could be made on fairly traditional grounds, such as cost reduction (usually by machines and systems replacing people) or the improved quality associated with better machine control. As information technology has matured, as the cost mix (hardware, software, and labor) associated with it has changed, and as its applications have broadened, it has become more difficult to comprehend the impact that an information system will have and, consequently, to judge whether or not an investment in the technology will pay off (CSTB, 1994a; Kaplan, 1986, 1989). The benefits of better understanding would extend to organizations of all sizes.
>
> It has proven particularly difficult to justify the costs of providing the infrastructure required to maintain and enhance information technology. There is no straightforward relationship between information services and the yield or quality of the finished product. Consequently, justifications for the infrastructure have to be based on more speculative, qualitative rather than quantitative, arguments. Better metrics are needed for measuring the value of support systems. These could also contribute to better cost management. Recent efforts to develop activity-based costing may be helpful to manufacturers in this context; additional research into applications of this approach in the manufacturing context may be useful.
>
> As companies attempt to become more agile, reduce the time required to respond to market needs, and explore the merits of "virtual enterprises," they will need tools to help determine what should be measured. How, for example, should one translate the metrics of agility into realistic financial terms? How can one measure the costs at organizational boundaries and judge whether information technology will improve an operation enough to justify the costs of the system? The speed with which enterprise-wide systems will be deployed will, to a large degree, depend on how well companies are able to understand these issues. That understanding would benefit from research into "exchange rates" between agility and financial metrics as well as case studies of enterprises that have implemented complex information systems.

desirability. For this reason, the committee is drawn strongly to the notion of an advanced long-range technology demonstration (ALRTD).[8]

ALRTDs focus on a variety of needs, covering both technical and cultural issues. ALRTDs are intended to provide an active means for transferring promising research into practice and move technological concepts closer to business

[8] An advanced long-range technology demonstration differs from the Advanced Technology Demonstration program of the National Institute of Standards and Technology in that the latter is focused primarily on demonstrations of technology just before it is mature enough to support product development. The technology associated with an ALRTD would generally be at an earlier stage of development at the time the ALRTD was proposed.

opportunities; connect researchers to problems, reveal knowledge gaps, and test research ideas; and bridge a variety of "believability gaps" that keep new ideas from being tried in the real world. Successful ALRTDs would be vehicles that could help to change manufacturing management skeptics into supporters and believers without making them suffer through a painful process of education within their own factory domain, and they would help answer the question, Is there a performance benefit to making changes on the factory floor?

For a candidate ALRTD to be viable and effective, it must be focused and bounded; Box 7.3 provides a number of areas in which ALRTDs might prove helpful. Since some investment in technology will almost certainly be required (in some cases, as much as that required for a pilot manufacturing plant), it must be funded well beyond the level of technical research grants. So that its success (or lack thereof) can be determined, quantitative goals or metrics that can serve as yardsticks should be established in advance. Moreover, each ALRTD should have core nontechnological content, such as a specific business activity, so that the technology will be the means to a clearly defined end rather than an end unto itself. Finally, each ALRTD should have some growth or follow-on potential and should not be a dead end.

Each ALRTD proposed should be based on a vision of what will be possible if the ALRTD is implemented, including what knowledge will be obtained if the ALRTD is successful. It should have available to it the necessary physical and information infrastructure, including other ALRTDs that could be under way at the same time or should have been completed first. It should clearly state the capabilities that will be demonstrated. Finally, it should describe expected consequences of a successful ALRTD that go beyond the explicit things to be demonstrated.

Appendix C describes several possible ALRTDs that are consistent with this description.

STIMULATING THE ADOPTION OF OPEN STANDARDS

As discussed in previous chapters, the lack of standards is a major inhibitor to the use of existing information technology for manufacturing. From computer-aided product design tools to real-time machine controllers, the lack of standards in various forms prevents data interchange, increases training costs, and impedes would-be product developers who have to build basic services into their products from scratch. Smaller manufacturers, especially, stand to benefit from the wide promulgation and acceptance of standards for interfaces and interconnections, because such standards are likely to lead to a much wider selection of information technology products for manufacturing applications (as more vendors will be able to develop products based on a common standard) and lower prices for such products (as vendors will be able to avoid the cost of developing unique solutions to common problems).

BOX 7.3 Areas for an Advanced Long-range Technology Demonstration

- Executive level
    Strategic analysis and evaluation of alternatives
    Strategy formulation
    Planning for implementation of strategies
- Finances
    Debit management
    Cash management
    Currency and commodities hedging
    Profitability
- Human resources
    Hiring metrics
    Team management
    Training
    Rewards
    Multinational teams
- Sales and marketing
    Gathering customer input
    Forecasting sales by product and geography
- Product design and development
    R&D, design, prototyping, and verification
    Design methodologies and management
    Computer-aided design tools
    Process design and verification
- Production
    Fabrication, assembly, logistics
    Distribution
    Inventory control
    Resource procurement and use planning
    Procurement, source qualification
    Maintenance and repair of facilities
- Legal concerns
    Alliances
    Contracts
    Safety
    Environment
    Liability

What prevents standards from being adopted more broadly? One answer is that the premature imposition of standards tends to freeze technological development and progress, and manufacturing is certainly one arena in which the use of information technology is relatively immature. But a second important barrier to adopting standards is not rooted in technical issues—a company trying to differentiate its products from those of other vendors may well choose to make them incompatible with others. Representing data in ways that facilitate interoperability may subject a vendor to infringement liability or require it to pay royalties.

Finally, once many competing products employ different representations and interfaces, a company may well resist adopting a standard proposed by another company for fear of ceding a commercial advantage to that company.

Previous chapters suggest the need for technical work on the development of open standards that will still allow technological progress. But there is a sociological dimension to standards as well—even when the underlying science and generic enabling technology are genuinely capable of supporting useful commercial applications, the vendor community needs to find some way to agree on standards that will expand the market for all.

Many problems need to be overcome if agreement is to be reached. For example, trends toward distributed intelligence, as well as the need to form alliances with other business entities (including outside design and engineering services as well as material and equipment suppliers), raise important knowledge management issues. How should companies determine what proprietary knowledge to share through distributed intelligence? What are ways to share such knowledge that protect knowledge assets and avoid disruptive interactions? How can commercial, off-the-shelf systems and technology best be combined with proprietary knowledge, perhaps in the form of software and knowledge bases? Such difficulties have already appeared in the very large scale integration applications-specific integrated circuit industry: vendors that own significant design libraries (for cells or "macrocells") are very reluctant to disclose detailed models of these cells to their customers, for fear that their customers' ability to do reliable simulations will hurt them competitively.

A second problem is how to achieve an appropriate balance between reducing the costs of variation and enabling rapid implementation of new technology—in short, the fundamental tension over whether and when to standardize. Research could help to establish criteria for determining when the benefits of restricting options by the adoption of standards outweigh the costs.

A third problem is how to generate widespread acceptance of and conformance to standards. Because of the large number of parties with vested interests in the particulars of any given standard, a careful and deliberative approach with wide representation is necessary. However, while wide representation is often a facilitator of consensus and acceptance, it is a factor that tends to slow down the consensus-forming process and also sometimes discourages the exploitation of technological opportunities that may spontaneously arise. In addition, once a standard is in place, a substantial amount of education is often necessary to familiarize potential users with its benefits and its technical implications.

DEVELOPING BETTER METRICS OF PERFORMANCE

Many metrics used today measure factory performance at high levels of aggregation that do not provide managers with insight into what is going on at the activity level. For example, metrics often used today include inventory turns (a measure of how rapidly inventory flows through the manufacturing system), defect rates (how often manufacturing problems occur), and on-time deliveries (the extent to which manufacturing schedules are met). These metrics are useful to signal the existence of problems somewhere within a manufacturing operation but cannot by themselves identify sources of the problems, which are affected by many factors.

Moreover, when the use of new technologies and approaches leads to new combinations of previously separate activities, many traditional measures of performance are problematic at best—and perhaps outright misleading or useless. For example, repetitive activities are much easier to measure than others, and performance metrics are often based on what is easy to measure. How does one use metrics based on repetition when the very point of information technology is to increase flexibility and the ability to cope with one-of-a-kind new situations and new sets of circumstances? Understanding in detail the impact of information technology is crucial for obtaining the best value, because only a detailed analysis can separate the impact of information technology from the effects of all of the other factors that may also impinge on overall performance.

Research is needed to develop good metrics that reflect the impact of information technology on manufacturing activities; a good beginning would be research on measuring the effect of information technology applications using existing metrics. Progress in this area will require both sophisticated technical understanding and a shared understanding within the community on the appropriateness of these metrics.

A related issue is that trustworthy cost-benefit analyses require reliable data. Obtaining meaningful data on costs is typically much more difficult than budget analysts anticipate. Consistent definitions are lacking, and typical cost categories and allocations are designed in accordance with conventional accounting rules rather than for tracking the performance of integrated business practices. New accounting categories based on the specifics of manufacturing may well be needed if an accurate picture is to be drawn of dollar flows in manufacturing.

DEVELOPING NEW MODELS FOR
ACCOUNTING AND VALUATION

The complexity of a large manufacturing business makes it very difficult for managers to identify operations that cost more than their value warrants. It is thus necessary to develop detailed models of a business that truly relate causes and effects, costs and benefits, and problems and solutions. Without such models, it will be difficult to convince anyone of the wisdom of taking certain courses of action, especially those that may be counter-intuitive.

Better management accounting systems are an integral part of these models. Timely activity-based accounting systems can provide explicit guidance regarding where costs can be trimmed and information on design parameters that demand resources.[9] With traditional accounting, the overhead costs of operating a plant are spread over all items produced in proportion to the volume of each type of item. But with activity-based accounting systems, the actual cost of small-volume production can be properly attributed and thus recovered.

Finally, generally accepted accounting principles that businesses use to audit their finances and operations are derived from a business philosophy in which capital expenditures (i.e., expenditures that relate to the long-term value of a company) are associated with buildings and pieces of equipment. Although this philosophy was appropriate in manufacturing 30 years ago, its relevance has decreased as the basis for competitive advantage has increasingly become the ability to exploit information more effectively. In particular, under old accounting principles the contribution of various types of knowledge (e.g., skilled personnel, software, and organizational structure) to the book value of a firm is essentially zero, even if that knowledge is the primary enabler of that firm's success.

It is admittedly difficult to place a book value on assets that are intangible and inscrutable to the layperson or that seemingly lack durability. Senior management is often legitimately concerned about the potential for misleading financial reporting as a product of estimating the value of intangible assets such as knowledge, and a variety of stakeholders (e.g., analysts, investors and owners, managers of organizations using software, managers of organizations producing software, tax authorities) are likely to have different judgments on what such assets are "truly" worth. Nonetheless, given the importance of knowledge to the future competitive environment, serious research to find valuation schemes that appropriately account for the contribution of knowledge to value is well worthwhile.[10]

EMPHASIZING EDUCATION, TRAINING, AND RETRAINING

The obstacles described above are among the factors motivating the current concern of management theorists and analysts with the "management of change." According to this perspective, successful enterprises of the future must change rapidly and frequently. Information, types of media, tools, processes, organizational structures, and so on will change, rendering knowledge and skills obsolete several times in an individual's career. Continuing education and training will have to inculcate adaptability and the expectation of change. These outcomes will be difficult to achieve because change provokes anxiety and stress. Research into techniques and tools to help enterprises and individuals understand and

---

[9] See, for example, Nanni et al. (1988).

[10] Similar concerns were raised for service industries in *Information Technology in the Service Society: A Twenty-first Century Lever* (CSTB, 1994a).

manage the process of change could facilitate the kinds of evolution of organizations and skills that many analysts anticipate.

Expectations for frequent change and greater employee autonomy imply that the manufacturing enterprise of the future will depend more on the knowledge and skill sets of each employee. Several needs follow from this assumption:

- Education of senior managers, to help them make informed decisions about using new, expensive, and perhaps risky information technology;
- Education and training of middle managers, to help them reduce barriers to trying new technology that may seem only to consume time and resources or imperil their own jobs;
- Education of designers, engineers, manufacturing engineers, and toolmakers to give them an increased technical base so that they can better understand and use computer technology and design relevant software;
- Training of factory workers, who are expected to use this new technology to improve their performance and that of their facility; and
- Retraining of factory workers whose skills are not compatible with the information technology paradigm and who may have to seek alternative jobs.

Even the term "frequent change" may understate the rapidity with which individuals will have to learn new things. Indeed, the volume of information that is relevant to successful individual performance is both so large and so rapidly changing that a learning model based on discrete phases of "retraining" may not be entirely sufficient. Another notion—what might be called "just-in-time learning"—is based on the idea that people should learn new skills and information as they need them to do what they must do. While such a notion has applicability that goes beyond manufacturing (indeed, it has ramifications for all of education), the manufacturing domain may be one that is particularly suitable for exploring the potential of this notion through intelligent tutoring systems, long-distance learning systems, and multimedia experiential learning tools.

Education issues are ubiquitous and topical; in addition, the renewal and currency of employee process- and product-specific knowledge have long been a major problem for all manufacturing enterprises. Needed are better means of delivery and better understanding of human issues such as motivation (or fear based on inadequate technical education and the resulting wish to fall back on intuition and experience).

Possible approaches to addressing the need for ongoing education include conferences and seminars for senior and middle managers; publication of educational materials; close collaboration among academicians, designers, engineers, and factory workers; and television and audio tape courses.

Computer aids—computer-aided instruction—also may be brought to bear on the education and training requirements of an enterprise, providing efficient and customized education that allows a person to learn quickly and selectively.

Instructional programs of the future will be more interactive and user-friendly than the lecture and fixed-sequence documentaries of today. Computer aids for learning may be built into new manufacturing technologies themselves, as well as installed in stand-alone workstations for educational purposes. Multimedia and virtual reality technology may hold promise for providing a flexible, socially acceptable, and nonthreatening interface for educational and skill-building programs; an NII and future home and factory systems offer the means to distribute and use such programs. The delivery systems will have to provide overview data as well as fundamental knowledge. NSF support of these technologies, perhaps through Education and Human Resources Directorate programs, could increase the skill base of the manufacturing sector, as well as of other sectors of society in general.

The challenge of providing new educational technology can easily be underestimated. But it is a complex, multifaceted task. The target population spans a wide range of educational backgrounds, capabilities, and needs. Some individuals will want a high-level overview of the material, and others will want in-depth courses with detailed practice sessions. The work force today spans multiple languages in the domestic United States, and specific topics will also be made available to people in other countries and from different cultures. Thus the content of educational programs and information is critical, the quality of presentation must meet modern expectations, and usability must be tunable to purpose.

Bibliography

Advanced Manufacturing Program, Office of Technology Commercialization, Technology Administration, U.S. Department of Commerce. 1989. *Economics, Management Policies and Financial Accounting Practices for a Shared Flexible Computer Integrated Manufacturing Facility.* U.S. Department of Commerce, Washington, D.C.

Amello, Gilbert F. 1993. "Managing the Integration of Semiconductor Manufacturing," remarks presented at International Semiconductor Manufacturing Sciences Symposium SEMICON/West on July 19.

Bartlett, Christopher A., and Sumantra Ghoshal. 1992. *Transnational Management: Text, Cases, and Readings in Cross-border Management.* Irwin, Homewood, Ill.

Bjorke, O. 1979. "Computer Aided Part Manufacturing," *Computers in Industry* 1:3-9.

Clinton, William J., and Albert Gore, Jr. 1993. "Remarks by the President and Vice President to Silicon Graphics Employees." Electronic mail press release dated February 22.

Clinton, William J., and Albert Gore, Jr. 1993. *Technology for America's Economic Growth, a New Direction to Build Economic Strength*, February 22.

Computer Science and Telecommunications Board (CSTB), National Research Council. 1992. *Computing the Future: A Broader Agenda for Computer Science and Engineering.* National Academy Press, Washington, D.C.

Computer Science and Telecommunications Board (CSTB) and Manufacturing Studies Board (MSB), National Research Council. 1993. *Information Technology and Manufacturing: A Preliminary Report on Research Needs.* National Academy Press, Washington, D.C.

Computer Science and Telecommunications Board (CSTB), National Research Council. 1993. *National Collaboratories: Applying Information Technology for Scientific Research.* National Academy Press, Washington, D.C.

Computer Science and Telecommunications Board (CSTB), National Research Council. 1994a. *Information Technology in the Service Society: A Twenty-first Century Lever.* National Academy Press, Washington, D.C.

Computer Science and Telecommunications Board (CSTB), National Research Council. 1994b. *Academic Careers for Experimental Computer Scientists and Engineers.* National Academy Press, Washington, D.C.

Council of Economic Advisors for the Joint Economic Committee. 1994. *Economic Indicators.* U.S. Government Printing Office, Washington, D.C.

Cowger, Gary L. 1991. Remarks made at "The Integrated Enterprise: Moving Beyond Technology CAM-I Conference," Orlando, Florida, September 25.

Cutkosky, Mark R., Robert S. Engelmore, Richard E. Fikes, Michael R. Genesereth, Thomas R. Gruber, William S. Mark, Jay M. Tenebaum, and Jay C. Weber. 1993. "PACT: An Experiment in Integrating Concurrent Engineering Systems," *IEEE Computer* 26(1):28-37.

Darlin, Damon. 1994. "Automating the Automators," *Forbes* 153(4):156-160.

Dertouzos, Michael L. 1989. *Made in America: Regaining the Productive Edge.* MIT Press, Cambridge, Mass.

Division of Information, Robotics, and Intelligent Systems, National Science Foundation. 1993. *Summary Report of the Intelligent Manufacturing Research Planning Meeting,* June 7-8, Washington, D.C.

Erisman, Albert M., and Kenneth W. Neves. 1987. "Advanced Computing for Manufacturing," *Scientific American* 257(4):163-169.

Eversheim, Walter. 1991. "Strategies and Tools Meeting Future Challenges in Manufacturing," paper presented at "International Conference on the Application of Manufacturing Technologies" held April 17-19 in Washington, D.C., and sponsored by Society of Manufacturing Engineers.

Friedman, Avner, James Glimm, and John Lavery. 1992. *The Mathematical and Computational Sciences in Emerging Manufacturing Technologies and Management Practices*, SIAM Reports on Issues in the Mathematical Sciences, Society for Industrial and Applied Mathematics, Philadelphia, Pa.

Genesereth, Michael R., and Nils J. Nilsson. 1987. *Logical Foundations of Artificial Intelligence.* Morgan Kaufmann, Los Altos, Calif.

Greenfeld, I., F.B. Hansen, and P.K. Wright. 1989. "Self-sustaining, Open-system Machine Tools," *Proceedings of the North American Manufacturing Research Institution* 17:281-292.

Hammer, Michael, and James Champy. 1993. *Re-Engineering the Corporation: A Manifesto for Business Revolution.* Harper-Collins, New York.

Harrington, Joseph. 1984. *Understanding the Manufacturing Process: Key to Successful CAD/ CAM Implementation.* M. Dekker, New York.

Holland, John H. 1985. "Properties of the Bucket Brigade Algorithm," *Proceedings of the International Conference on Genetic Algorithms and Their Applications,* J.J. Grefenstette (ed.). Carnegie Mellon University, Pittsburgh, Pa.

Initial Graphics Exchange Specification (IGES) Version 5.2. 1993. U.S. Pro, Fairfax, Va.

Interchem On-Line. 1993. "A Day in Your Not-too-distant Future?" Vol. 1, No. 1, pp. 2-3.

International Organization for Standardization (ISO). 1994. *Industrial Automation Systems and Integration—Product Data Representation and Exchange, Part 1: Overview and Fundamental Principles.* ISO 10303-1. ISO, Geneva.

Jaikumar, Ramchandran. 1988. "From Filing and Fitting to Flexible Manufacturing," Harvard Business School working paper No. 88-045, Boston, Mass. Reproduced in R. Jaikumar, "200 Years to CIM," *IEEE Spectrum* 30(9):26-27.

Kahaner, David K. (U.S. Office of Naval Research Asia). 1993. "A Survey and Discussion of CIM Projects in Japan." Electronic mail message dated January 13, 1993, to "Distribution" from kahaner@cs.titech.ac.jp.

Kaplan, Robert S. 1986. "Must CIM Be Justified by Faith Alone?" *Harvard Business Review* 64(2):87-95.

Kaplan, Robert S. 1989. "Management Accounting for Advanced Technological Environments," *Science* 245(4920):819-823.

Lardner, James F. 1986. "Computer Integrated Manufacturing and the Complexity Index," *The Bridge* 16(1):10-16.

Luss, Hanan, Moshe B. Rosenwein, and Elizabeth T. Wahls. 1990. "Integration of Planning and Execution: Final-Assembly Sequencing," *AT&T Technical Journal* 69(4):99-109.

Manufacturing Studies Board (MSB), National Research Council. 1988. *A Research Agenda for CIM.* National Academy Press, Washington, D.C.

Manufacturing Studies Board (MSB), National Research Council. 1991. *Improving Engineering Design: Designing for Competitive Advantage.* National Academy Press, Washington, D.C.

Marks, Peter. 1993. *15 Key Processes in the CASA/SME Wheel.* Draft copy.

Merchant, M.E. 1971. "Delphi-type Forecast of the Future of Production Engineering," *ICIRP Annals* 20(3):213-225.

Miller, Jeffrey G., and Thomas E. Vollmann. 1985. "The Hidden Factory," *Harvard Business Review* 63(5):142-150.

Moore, John. 1994. "Agencies, Vendors Split on IDEF3 Techniques," *Federal Computer Week,* August 1.

Mukherjee, Amar, and Jack Hilibrand (eds.). 1994. *New Paradigms for Manufacturing.* NSF 94-123. National Science Foundation, Washington, D.C.

Nanni, Alfred J., Jeffrey G. Miller, and Thomas E. Vollmann. 1988. "What Shall We Account For?" *Management Accounting* 69(7):42-48.

National Critical Technologies Panel. 1991. *Report of the National Critical Technologies Panel.* The National Critical Technologies Panel, Arlington, Va., March.

National Institute of Standards and Technology (NIST). 1994. *Putting the Information Infrastructure to Work: A Report of the Information Infrastructure Task Force Committee on Applications and Technology.* SP857. National Institute of Standards and Technology, Gaithersburg, Md., May.

National Research Council. 1994. *Virtual Reality: Scientific and Technological Challenges.* National Academy Press, Washington, D.C.

O'Lone, Richard G. 1991. "777 Revolutionizes Boeing Aircraft Development Process," *Aviation Week & Space Technology* 134(22):34-61.

Pan, Jeff Y.-C., Jay M. Tenenbaum, and Jay Glicksman. 1989. "A Framework for Knowledge-based Computer-Integrated Manufacturing," *IEEE Transactions on Semiconductor Manufacturing* 2(2):33-46.

Pan, Jeff Y.-C., and Jay M. Tenenbaum. 1991. "An Intelligent Agent Framework for Enterprise Integration," *IEEE Transactions on Systems, Man, and Cybernetics: Special Issue on Distributed Artificial Intelligence,* September.

Port, Otis. 1991. "'This Is What the U.S. Must Do to Stay Competitive,'" *Business Week* No. 3244, December 16, p. 92.

Porter, Michael E. 1992. *Capital Choices: Changing the Way America Invests in Industry.* Council on Competitiveness and Harvard Business School, Washington, D.C., June.

Sadeh, Norman. 1993. *Micro-Opportunistic Scheduling: The Micro-Boss Factory Scheduler.* The Robotics Institute, Carnegie Mellon University, Pittsburgh, Pa.

Shiraishi, M. 1988. "Scope of In-Process Measurement, Monitoring and Control Techniques in Machining Processes," Part 1—"In-Process Techniques for Tools," *Precision Engineering* 10(4):179-189.

Shiraishi, M. 1989a. "Scope of In-Process Measurement, Monitoring and Control Techniques in Machining Processes," Part 2—"In-Process Techniques for Workpieces," *Precision Engineering* 11(1):27-37.

Shiraishi, M. 1989b. "Scope of In-Process Measurement, Monitoring and Control Techniques in Machining Processes," Part 3—"In-Process Techniques for Cutting Processes and Machine Tools," *Precision Engineering* 11(1):39-47.

Siewiorek, Daniel P. 1992. "Rapid Prototyping: The Design Process, Tools, and Fabrication," presented at "Information Technology and Manufacturing: A Workshop," National Science Foundation, Washington, D.C., May 5-6.

Sousa, Louis J. 1993. "Evidence of a New, Value-added Materials Paradigm," *Journal of the Minerals, Metals and Materials Society* 45(6):9-11.

Spur, Günter (ed.). 1990. *Production Technology Centre, Berlin,* Fraunhofer-Institut für Produktionsanlagen und Konstruktionstechnik, Berlin.

Tenenbaum, Jay M., and Rick Dove. 1992. "Agile Software for Intelligent Manufacturing," presented at "Information Technology and Manufacturing: A Workshop," National Science Foundation, Washington, D.C., May 5-6.

Tenenbaum, J.M., R. Smith, A.M. Schiffman, A. Cavalli, and M. Fox. 1991. *The MCC Enterprise Integration Program: A Prospectus.* Preliminary report, June 4.

The Economist. 1994. "Manufacturing Technology: On the Cutting Edge," Vol. 330, No. 7853, pp. 3-18.

Trobel, Russ, and Andy Johnson. 1993. "Pocket Pagers in Lots of One," *IEEE Spectrum* 30(9):29-32.

U.S. Air Force Manufacturing Technology Directorate. 1991. *Automated Airframe Assembly Program.* NB91-51. Northrop Corporation Aircraft Division and Wright Research and Development Center, July.

U.S. Air Force Manufacturing Technology Directorate. 1993. *Producibility Methodology and Tool Set (PMATS) Initiative: Concept Development and Program Strategy.* A white paper "final draft" dated May 1, Wright Research and Development Center, Cincinnati.

U.S. Department of Commerce, Economics and Statistics Administration, Bureau of Economic Analysis. 1994. *Survey of Current Business.* U.S. Government Printing Office, Washington, D.C., April.

Wall Street Journal. 1994. "European Car Recycling Accord," April 29, p. A9E.

Ward, Elaine. ND. *IDEF.* MITRE-Washington Software Engineering Center, MITRE Corporation, McLean, Va.

Weiss, L.E., F.B. Prinz, D.A. Adams, and D.P. Siewiorek. 1992. "Thermal Spray Shape Deposition," *Journal of Thermal Spray Technology* 1(3):231-237.

Wysk, Richard. 1992. "Integration Requirements for Intelligent Manufacturing," presented at "Information Technology and Manufacturing: A Workshop," National Science Foundation, Washington, D.C., May 5-6.

Appendixes

A

List of Contributors

Suzanne Barber, University of Texas at Austin
Craig Barrett, Intel Corporation
Steve Benson, Thesis Inc.
John Birchak, Intel Corporation
Robert E. Boykin, Consortium for Advanced Manufacturing International
Douglas F. Busch, Intel Corporation
Mark R. Cutkosky, Stanford University
Alan De Pennington, University of Leeds
Richard Dove, Paradigm Shift International
Ken Evans, Réseautique
David R. Ferguson, Boeing Computer Services
David Files, Eastman Kodak
Donald A. Jenkins, Allied Signal
Reuben S. Jones, Softech Inc.
Robert E. Joy, Northrop Corporation
Robert Kaplan, Harvard Business School
Karl Kempf, Intel Corporation
John K. Korah, Electronic Data Systems Corporation
Alison McKay, University of Leeds
Greg McLaughlin, Sun Microsystems Laboratories Inc.
Gordon Moore, Intel Corporation
Richard E. Morley, Flavors Technology Inc.
Victor Muglia, Caterpillar Solar Turbines Inc.
Mark Pearson, University of Leeds

George Pfeil, Ford Motor Company
Jan Pounds, Minnesota Technology Inc.
Michael Pratt, Rensselaer Polytechnic Institute
Ari Requicha, University of Southern California
Charles M. Savage, Digital Equipment Corporation
Warren L. Shrensker, General Electric (retired)
Bruce Sohn, Intel Corporation
William J. Spencer, Sematech Inc.
N.S. Sridharan, Intel Corporation
John M. Swihart, National Center for Advanced Technologies
George Thompson, Johnson & Johnson
David Ullman, Oregon State University
Ernest O. Vahala, General Motors North American Truck Platforms
Herbert B. Voelcker, Cornell University
Frank Wilde, Collaborative Technologies Corporation

B

Site Visit to Romeo Engine Plant
March 23, 1994

A FACTORY DESIGNED TO USE INFORMATION TECHNOLOGY

Ford's Romeo engine plant is its newest, having started production in the summer of 1990. Its 750 employees and approximately $1 billion, 3-million-square-foot physical plant make a wide variety of V-8 engines totalling about 2,400 per day. The plant buys about 70 percent (by value) of what it ships and devotes most of its energy to machining engine blocks, pistons, connecting rods, and crank shafts and then assembling everything into a complete tested engine ready to install in a car. Plant manager George Pfeil believes that the plant could not be operated without information technology.

Information technology (IT) exerts its effects in two ways: by empowering people and by organizing the flow of people and parts. This is a very clean and highly automated plant with a large number of robots performing assembly and thousands of feet of conveyor lines moving parts through a series of machining operations. IT permits a relatively small number of people to keep all these machines running and all the complex operations on track. For example, one line for machining engine blocks is 2,500 feet long and is operated by 8 people per shift. In older plants without extensive IT, 30 people would be needed.

When the plant was designed in the late 1980s, IT was a central part of the design. The plant was equipped with a network of what Pfeil calls "white courtesy phones," which are actually input/output consoles on which all production information is available to all employees. Similarly, the controllers of the machines were wired into these consoles as well as into a pager system so that machine stoppages could be recorded and the responsible person paged automati-

165

cally. Even though several U.S. and foreign companies built the machining and assembly lines, all used a common Programmable Logic Control (PLC) system specified by Ford and linked their machines to it.

There are no first-line supervisors, and shop floor personnel are expected to answer the pagers and keep their machines operating. In a typical day there are between 12,000 and 16,000 manual and automatic pages. In this way IT enables the shop floor personnel to take command of the machinery, keep it running, and make the necessary decisions to do so. The United Auto Workers Union is fully involved and supports this method of operation.

Clearly, such a system could not have been overlaid onto a previously designed factory. Designers thought through how this factory would be linked by IT as part of their operating philosophy and human relations strategy, and IT was up and running the day production started. Pfeil says in retrospect that it was not easy explaining to upper management why this kind of plant needed so much IT, but the designers did it anyway. Today the benefits can be accounted for both qualitatively and quantitatively:

• A small number of spirited union employees can turn out a very large number of engines; it is a "world-class operation."
• Statistics on "things gone wrong per thousand [engines]" in the first 6 months after the launch of a new engine reveal an average of over 150 on typical launches in the 1980s versus 37 on the first Romeo launch in 1991 and 41 on the second in 1993.

HOW INFORMATION TECHNOLOGY IS USED

In addition to the white courtesy phones, IT is used in several other ways at the Romeo engine plant. For example:

• In the tool crib, an electronic display indicates how to set up each tool and adjust it for proper alignment prior to installing it on a machine.
• Each time an engine must be diverted from the assembly line for minor rework, the operator records the cause and the remedy; that engine is then tracked in the warranty system after it is sold to see if any unusual problems arise.
• A bar code on each engine tells the cold test machine what equipment is on the engine so that the correct test can be used.
• When the cylinder head is bolted to the block, the bolt torque is sensed and recorded.
• A Machine Monitoring System keeps track of equipment status. According to the plant newspaper, 598 machines are monitored and 15,000 items are recorded each minute. Each time a machine, robot, or conveyor stops, the PLC deduces the cause from sensors and the state of the control logic; the cause is recorded and accumulated with other causes to form the basis for the reports

described below. Pfeil notes that accurate data of this kind are essential and cannot be gathered by people, who tend not to notice short stoppages (the vast majority) and cannot always be present when breakdowns occur to note their cause.

• Each week performance data on each section of the plant accumulated by the PLCs are collected and packaged into Pareto charts that identify bottleneck stations and the top 10 reasons for their failing to keep up; recent data are compared to 7-week trends so that plant personnel can tell if they are making progress in improving uptime; these reports are available to all employees on the white courtesy phone consoles. The bottleneck is identified by a formula that compares theoretical with actual output, adjusted for faults on other machines. Pfeil decided long ago that people are poor judges of which machine is the bottleneck.

Scheduling is currently done using manufacturing resources planning (MRP-II), but Pfeil is searching for better methods that take account of constraints. A mathematician by training, he is familiar with such methods as linear programming and OPT (a commercial factory scheduling program that takes constraints like machine capacity into account). Pfeil notes that any success in *Kaizen* (continuous improvement) will "cause the bottleneck to move around." That is, as one machine's capacity is increased or a breakdown is fixed, the bottleneck will by definition shift to the next slowest or most constrained machine. Scheduling software must be capable of being easily reprogrammed to keep up, or else the schedules will rapidly become useless. Currently reprogramming is not possible, or it takes so long (typically hours) that the situation has changed already.

Frank Keene, Cast Iron Block Machining Line manager, notes that by 1996 there will be a step increase in production complexity as the California emission standards begin to take effect. Managing all the machines, parts, and people will become increasingly difficult. An important problem is reconfiguring the plant to make a different-model engine. This requires shifting people from one area to another, an action that can take 30 to 45 minutes. Right now he has no accurate way of predicting how many people will be needed for each type or how long it will take to move them. Thus there is some waste in plant operations that must be eliminated in order to absorb the expected increase in complexity.

TRAINING FOR USE OF INFORMATION TECHNOLOGY

Training is essential to the success of the Romeo plant. It is such a strong part of the plant's culture that Pfeil teaches one of the courses, called Productivity 201. Training starts with operation of the white courtesy phones and proceeds to sensitize the employees to the importance of keeping the plant running. Basic to this is building an appreciation of the value of the data that can be obtained via the white courtesy phones. These data include the Pareto charts and other data. Productivity 201 and other courses utilize a simulation of a typical line written in

Witness (a commercial package for simulating discrete events) by a professor from the University of Detroit. A sample class exercise asks the students to look at the Pareto chart, identify the causes of downtime, and schedule preventive maintenance. When the simulation is complete, the students can see if productivity has increased. This and similar exercises help teams to develop better problem-solving methods and improve team dynamics.

Employees do not use this system to help them decide how to schedule actual preventive maintenance. However, Pfeil and his staff use it to plan production a week ahead as well as to see if their plans produce the anticipated production of all the required varieties of engines. Although basic training of new employees is done using company-wide course material, Romeo found that it had to generate its own advanced training curricula. Topics include such business issues as preventive maintenance (discussed above), quality, basic finance, and productivity. Interestingly, although other plants are copying the approach of using the white courtesy phones, Romeo's training materials are not being widely used in other Ford plants.

WHAT NEEDS TO BE IMPROVED

In addition to shortcomings in scheduling algorithms, Pfeil notes that user interfaces are too cumbersome and take too long to learn. They are also too text-oriented. He thinks an interface like that on the Macintosh would be better. (At Project Alpha, Ford's technology transfer operation, David Wood said that they need better "information ergonomics," meaning not only better interfaces but also better tools for turning data into useful information so that people know what to do.)

More broadly, Pfeil feels that advanced factories face huge "sociology problems." The plant is large and the company is even larger. It is difficult for one person to know what to do so that the whole thing improves. Currently, U.S. management uses "incentives" to direct employees' behavior. The problem is that incentives are substitutes for really knowing what action is best. Furthermore, current incentives for hourly employees are not always consistent with those of salaried employees. It would be better if IT could be used to inform all employees about how they were doing and how their personal returns could be improved while those of the company also improved. The white courtesy phones are a step in that direction, but only a first step. In particular, more needs to be learned about how teams operate, how they process data and reach decisions, and how best to present data to them. Productivity 201 demonstrates an intermediate version of what should be the better methods of the future. Since the Romeo plant will face increasing complexity within a very few years, such improvements are needed soon.

C

Illustrative Advanced Long-range Technology Demonstrations

The projects described below are intended to illustrate the types of advanced long-range technology demonstrations (ALRTDs) suggested in Chapter 7. These sketches omit many important issues, such as project size, participants (universities, companies), duration, budget, and funding mechanism; intellectual property concerns; and so on. However, given the broad scope of some of these projects, support of ALRTDs by mission-oriented agencies as well as research agencies would be desirable. Each project is described briefly in terms of its vision, the infrastructure it would require, the objectives of the demonstration, and its significance.

A MANUFACTURING ENTERPRISE MODELING MANAGEMENT SYSTEM

Vision

An enterprise faces many problems in the course of doing business. To develop insight into these problems and to evaluate different courses of action, decision makers often rely on computer-based models of these problems as they apply to the enterprise. Today, such models are often built "from scratch," because the problems span a set of organizational or functional boundaries that differ from those related to previous problems that have been modeled. However, the model of the problem at hand still requires access to most of the same databases that have been used previously; what is different most often relates to data form, aggregation, granularity, period, frequency, precision, and so on. A manufacturing enterprise modeling management system (MEMMS) would manage the reconfiguration of databases and existing models to account for these specific needs (Table C.1).

Infrastructure

A MEMMS would require a complete, modern digital communications system, most likely including a wide area network, a local area network with the necessary bandwidth, network management software, network servers, and individual man-machine interfaces; also included would be bridges into databases external to the enterprise. The enterprise would require a standard language and dynamic data dictionary, a distributed database management system, and digitally accessible archival systems. Models of different aspects of the business would have to be stable enough to allow the enterprise itself to evolve.

Objectives of the Demonstration

• To show that a MEMMS would provide direct, tangible, bottom-line benefits to the enterprise in flexibility, response time, and the complexity of problems that can be tackled. Problems could be addressed quickly and efficiently when an organization had completed the up-front work of standards, common language, data elements definition, database design, and modernization of the archival knowledge bases.

• To show that an appropriate methodology for model construction within an enterprise involving techniques of modular design, scaling, enterprise model hierarchy, and modeling software architecture would allow individual models and databases to be reconfigurable and reusable for new problems.

• To show that a MEMMS is possible that would enable employees in all functions within an enterprise to manage reconfiguration of a set of models for meaningful analysis and solution of problems facing manufacturing enterprises today. A corollary is that a software interface for a MEMMS is possible that does not require highly specialized training in its operation.

Significance

The MEMMS would provide a common language and analysis tools that would accelerate decision making but provide all players with access to the best and most current data and tools. Organizations would thus be able to focus more on the enterprise problems and less on the task of building new models. The interdependence of disparate business functions within the organization would become clear, and integration of all activities would accelerate. These changes would allow organizations to become more focused, flexible, and responsive to the marketplace.

TABLE C.1 A Manufacturing Enterprise Modeling Management System

	Example Models	Example Databases
Research and development	Process, material, and product concept models	Material properties data, performance data, boundary data
Product design	Process and product models—design, fabrication cost	Material performance data, production costs, design history library
Production	Process and product models	Data on volume, mix, cost, materials, maintenance, uptime
Supplier management	Product and process models, production and throughput models	Statistical process data, status and cost data, production data
Logistics	Shop floor production planning and throughput models	Data on volume, condition of paths of material flow, shipping schedules
Financial management	Financial models—production, debt structuring, investment	Production costs, financing costs, external interest rates
Human resources management	Cost-per-employee, succession, and training models	Compensation data, employee ratings, government regulations
Sales and marketing	Price-cost-market share models, sales forecasting models	Customer needs; costs to produce, ship, and advertise; data on customer needs, preferences
Business strategy and planning	Dynamic strategic analysis models, scenario models	Macroeconomic data; data on markets, competitors, costs, exchange rates
Alliance-joint venture management	Process and product models, financial models, distribution and flow models	Data on production, processes, products, costs, revenues, technology used
External stakeholder management	Life-cycle models, legislative and regulatory models	Data from the Environmental Protection Agency, Occupational Safety and Health Administration, legislative records, newspapers

NOTE: The manufacturing enterprise modeling management system would connect models and internal databases through a network and would access external databases as necessary.

OPERATION OF A MACHINE SHOP BROKERAGE SYSTEM

Vision

A machine shop brokerage clearinghouse would be given responsibility for on-line brokerage policies and procedures for participating customers, permitting their machining needs to be matched to the available machining capacity of a certain set of suppliers. The broker's responsibilities would include the establishment of machine shop certification and qualification, bidding procedures, request-for-quote procedures, arbitration of disputes, payment, and so on. The clearinghouse would solicit suppliers and advertise for customers; participating customers and suppliers would be connected by network in a given geographic region.

Infrastructure

The region served would require a wide area network and easy access of customers to the network. Suppliers and customers would have to be skilled enough to handle such things as computer-aided design data exchange, electronic data interchange requests for quotes and invoicing, electronic funds transfer payments, and so on. Suppliers would have to be sophisticated enough to handle statistical process and quality control techniques, provide records of employees' training and qualifications, and use computerized production order tracking and status reporting.

Objectives of the Demonstration

• To show that a set of policies and procedures could be developed for the purpose of electronic brokering of machine-shop-type work tasks and assurance that this type of business could be supported by existing technology.

• To show that a clearinghouse of this nature would enable customers to find services of assured quality and dependability and at reasonable cost.

• To show that technical and business issues between customers and suppliers (answering questions, modifying the design) could be handled over the network.

Consequences

The traditional business practice in which companies deal with a small number of familiar suppliers would change. Brokering would allow an enterprise to become more flexible and cost-effective by providing a menu of certified suppliers with differing availability, cost, skill sets, and quantity attributes.

BROKERAGE FOR NATIONAL PROTOTYPING FACILITIES

Vision

Prototyping technology is being developed in a number of U.S. universities, private industries, and national laboratories. To demostrate their capabilities and to increase customer awareness of and access to these facilities, a national brokerage institute would take responsibilty for on-line policy and procedures such as the establishment of prototyping site certification and qualification, bidding procedures, request-for-quote procedures, arbitration of disputes, payment, and so on. The institute would solicit members and advertise for customers. If it were successful, the flow of business into these facilities would help to amortize their high cost.

Infrastructure

The institute would require a wide area network and easy access of customers to the network. Suppliers and customers would have to be advanced enough to handle computer-aided design data exchange, electronic data interchange requests for quotes and invoicing, electronic funds transfer payments, and so on. Suppliers would have to be sophisticated enough to handle statistical process and quality control techniques, provide records of employees' training and qualifications, and use computerized production order tracking and status reporting.

Objectives of the Demonstration

• To show that a set of policies and procedures could be developed for the purpose of electronic brokering of prototyping work tasks and that this type of business could be supported by existing technology.

• To raise the visibility of limitations in software, standards, and the prototyping processes and also pinpoint industries in which large markets for this type of work exist. This should spur further research and development along the lines that industry really needs.

Consequences

A prototyping brokerage would expand access to and knowledge of prototyping processes and capabilities. Prototyping decreases design time by validating design decisions or providing early notice of design errors.

REAL-TIME SALES FORECASTING

Vision

A company should be able to locate and query financial, economic, and demographic databases and predict sales of one of its products by regional market. With such information in hand, it would then notify its supplier base of its anticipated needs for a defined time period in the future.

Infrastructure

A real-time sales forecasting system would require databases, search techniques, and modeling methods for extrapolating information that is not available directly. Of particular relevance are national and regional econometric data on and models of consumer behavior at the present time; data on stocks of finished goods unsold in a company's own pipeline plus estimates of stocks in competitors' pipelines extrapolated from other economic data; methods of determining demand for the company's products based on demand from allied economic fields (e.g., demand for electric sockets based on demand for wire, or demand for paint and varnish based on demand for lumber); and access to statistics from home shopping systems.

Objectives of the Demonstration

- To show that a company could find needed data.
- To show that a company could construct the models it needed if models did not exist, or that it could find experts to help construct models.
- To show that a company could obtain valid sales forecasts more rapidly with information technology than by former means, and/or that it could obtain more accurate forecasts than before.
- To show that a company could integrate these forecasts advantageously into dealings with its suppliers or business partners.

Consequences

Normal business practices would change greatly because each company as well as its competitors would have better forecasts. Thus each firm would have to add its own capabilities to what is available to everyone on the network. Relationships with suppliers would become tighter, ordering patterns would change to ever-smaller orders spaced more closely together, and inventories would become smaller, saving working capital.